THE LITTLE GUIDES

MAMMALS

THE LITTLE GUIDES

MAMMALS

CONSULTANT EDITOR
Dr. George McKay

FEDERAL STREET PRESS

This 2001 edition published by
Federal Street Press
A Division of Merriam-Webster, Incorporated
PO Box 281
Springfield, MA 01102

Federal Street Press books are available for bulk purchase for sales promotion and premium use. For details write the manager of special sales, Federal Street Press, PO Box 281 Springfield, MA 01102

Publisher: Sheena Coupe
Associate Publisher: Lynn Humphries
Art Director: Sue Burk
Senior Designer: Kylie Mulquin
Editorial Coordinators: Sarah Anderson, Tracey Gibson
Production Manager: Helen Creeke
Production Assistant: Kylie Lawson
Business Manager: Emily Jahn
Vice President International Sales: Stuart Laurence

Project Editor: Mary Halbmeyer
Designer: Moyna Smeaton
Consultant Editor: Dr. George McKay

ISBN 1-892859-24-6

Color reproduction by Colourscan Co Pte Ltd
Printed by LeeFung-Asco Printers
Printed in China

01 02 03 04 05 5 4 3 2 1

CONTENTS

PART ONE

THE WORLD OF MAMMALS

PART TWO

KINDS OF MAMMALS

PART ONE

THE WORLD OF MAMMALS

WHAT IS A MAMMAL?

Mammals are a class of animals with around 4,300 species. Compared with the three to 10 million possible animal life forms, mammals comprise a relatively small percentage, but their sheer diversity is astonishing. To help us understand this diversity, the animal kingdom is classified into groups, based on the evolutionary relationships between species.

Characteristics of mammals

Generally, mammals are warm-blooded animals that suckle their young on milk, have a body covering of hair or fur, prominent external ears, and a mouth armed with teeth. Not all mammals share all these characteristics however, and for scientists the single feature that differentiates the group is that the dentary bone of the lower jaw articulates directly with the skull. Unlike reptiles, the upper jaw is firmly attached to the rest of the skull.

SOCIAL BEHAVIOR OF LIONS

A pride usually consists of five to six adult females that are all related, unrelated adult males, and any cubs. Females are the backbone of lion society: they are the hunters, cub rearers, property owners and defenders. Cubs can suckle from any female in the pride and usually remain with their mother for two years.

TYPICAL MAMMAL HAIR

Hair is a feature unique to mammals. The variety of patterns shown by the medulla and cuticle cells can be used to identify species of mammals: the spines of the echidna, for instance. Some hairs are continually replaced; others periodically (annual molt); others die but are replaced by a neighboring follicle.

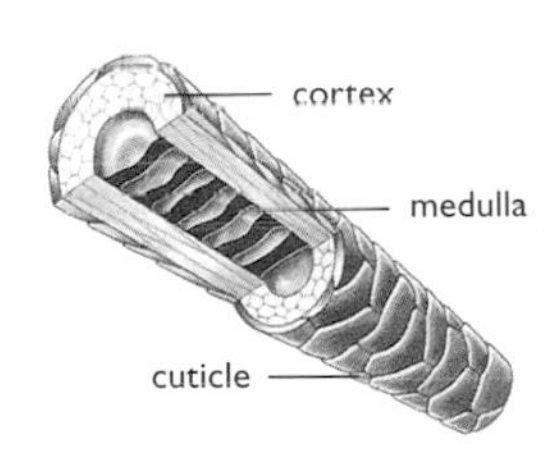

COMPLETE FOOD

Mammals are named after the mammary glands, found only in this class of animals. Milk provides the total nutrients required, changing in composition to match the needs of the developing young.

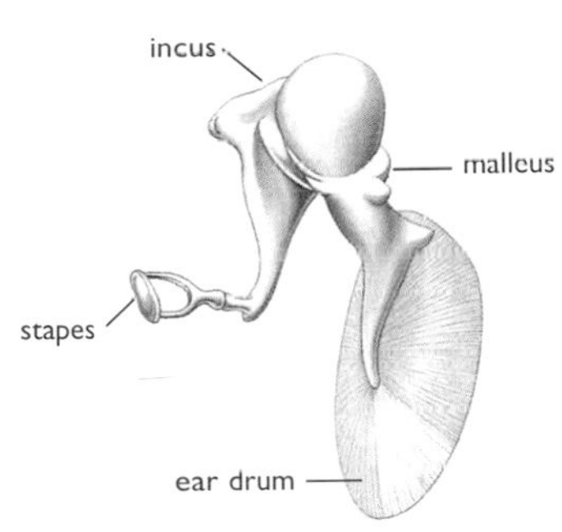

MAMMAL HEARING

Two of the three bones in the mammalian ear, the malleus and incus, have evolved from controlling the jaw hinge, as in reptiles, to assisting and amplifying hearing. In primitive mammals, this middle ear structure lies outside the skull. It is protected by a bony covering, the bulla, in more specialized forms.

CLASS MAMMALIA

Two major divisions of mammals are recognized. The Prototheria includes the echidnas of Australia and New Guinea, and the Australian platypus. These monotremes are the most ancient of mammals and reproduce by laying soft-shelled eggs. The other subclass, the Theria, give birth to live young. They are further divided into the Metatheria, or marsupials, whose young are born in an embryonic state, and complete their development in a pouch or fold on the mother's abdomen; and Eutheria, or placental mammals, whose young develop to a relatively advanced stage before birth.

Orders and Families There are around 20 orders of mammals that contain families sharing major characteristics. Most orders contain a number of families , though some contain a single, unique family. Only the aardvark has a single species in the order. This book is arranged, as in classification, by orders of mammals, with representatives of families forming the species descriptions.

THE WORLD OF MAMMALS

THE ORIGINS OF MAMMALS

Vertebrate history began around 510 million years ago (mya) with the first jawless fishes of the warm and wet Ordovician period, when the Earth's continents were fragments lined up along the equator. As the landmasses continued to move, the first amphibians appeared 400 mya. Over 200 million years later, mammal-like reptiles began to diversify, and by the late Jurassic, tiny mouse-like creatures diversified. Monotreme fossils have been dated to 130 mya, and marsupials and placentals appeared around 95 mya. Primates appeared 65 mya with another great diversification of mammals, and the forerunners of humans arose a mere 2 mya.

MAMMAL ANCESTORS

Three hundred million years ago (mya), the land was populated by primitive amphibians and reptiles, living in and around extensive tropical swamps. Some of the larger fossil reptiles had modifications in the hind part of their skulls, which are similar in form in mammals. Mammals may be considered "improved" reptiles.

DIMETRODON
One of the earliest and most primitive of the "mammal-like reptiles," or synapsids, *Dimetrodon* grew to about 10 feet (3 m) in length. It inhabited parts of what is now North America 300 mya. This finback reptile is a member of the group from which mammals evolved.

Synapsids *Dimetrodon* and her kin shared features with both mammals and reptiles, so that the earliest "mammal-like reptiles" are not definitively assigned to either group. They had a pair of windows, or temporal fenestrae, in their skulls, like mammals, but were coldblooded, like reptiles, and had simple teeth with weak jaws and clumsy, sprawling limbs.

CYNOGNATHUS

This carnivorous cynodont was an advanced version of the "mammal-like reptiles." The size of a badger, it had powerful jaws and much more mammalian teeth and bones. It is a matter of speculation whether these animals were warmblooded.

Therapsids The *Therapsida* evolved from *Dimetrodon*-like ancestors. They developed more powerful jaw-closing muscles and more elaborate dentitions, such as canines, or even horny beaks. They possessed longer, slender limbs giving greater agility and speed, and probably became warmblooded and larger-brained.

Cynodonts An advanced therapsid evolved several other distinctly mammalian characteristics. Teeth with several cusps enabled shearing and crushing, with jaw muscles that gave greater biting power. The lower jawbone enlarged to carry the attachments of these muscles.Other bones of the lower jaw became reduced and adopted a new function of transmitting sound waves. In mammals, these bones have lost contact with the jaw altogether and have become the sound-conducting mechanism of the middle ear.

MAMMAL ANCESTORS

- Synapsids: Middle Carboniferous (300 mya)
- Therapsida: Late Permian (255 mya)
- Cynodontia: Early Triassic (235 [illegible]
- Morganucodontids: Late Triassic (215 mya)

MAMMAL BEGINNINGS

Sometime around the end of the Triassic period, about 195 mya, the first fossils with a fully developed jaw hinge are found. They are the earliest true mammals, evolving from the advanced cynodonts. However, they were very primitive, still with other small lower jaw bones and a hinge bone attached to the lower jaw.

MEGAZOSTRODON
This tiny insect-eating animal that looked like a shrew is one of the oldest known mammals. It lived in Africa 220 mya.

Mesozoic radiation The earliest mammals were the start of an evolutionary radiation into several groups. The defining step was the development of a more complex tooth: the tribosphenic, or triangular, molar. The egg-laying monotremes were an early divergence from this group. By 80 mya, two fossil tribosphenic mammals could be distinguished: marsupials and placentals.

UINTATHERIUM The first large, browsing herbivores arose around 60 mya. The strange-looking uintatheres of North America were comparable in size to present-day rhinoceroses.

ANCIENT TIMES (MYA)

MESOZOIC ERA 245 TO 65
Triassic period 245 to 208
Jurassic period 208 to 145
Cretaceous period 145 to 65

CAINOZOIC ERA 65 MILLION YEARS AGO TO PRESENT

TERTIARY SUBERA 65 TO 1.64
Paleogene period 65 to 23, comprises 3 epochs
Paleocene epoch 65 to 56
Eocene epoch 56 to 35
Oligocene epoch 35 to 23.3
Neogene period 23.3 to 1.64, comprises 2 epochs
Miocene epoch 23.3 to 5.2
Pliocene epoch 5.2 to 1.64

QUATERNARY SUBERA and Pleistogene period 1.64 to present, comprises 2 epochs
Pleistocene epoch 1.64 million to 10,000 years ago
Holocene epoch 10,000 years ago to present

MORGANUCODONTIDS
Another example of the first mammals, thought to have been adapted to a nocturnal, carnivorous hunting existence. Its sharp teeth were used to capture and chew insects, and quite probably other terrestrial invertebrate prey.

Worldwide distribution
Morganucodontids are the best known of the earliest true mammals, found in Europe, South Africa, North America and China, which indicates a worldwide distribution. They were very small, with an overall body length of 5 inches (12 cm). Their teeth were sharp and multi-cusped and they seemed to have had an enhanced sense of hearing and smell.

MAMMAL DIVERSIFICATION

The Mesozoic mammals were successful in their radiation into different groups worldwide. Without exception, however, they remained very small. Perhaps they could not compete with large dinosaurs in their terrestrial habitats and were restricted to a nocturnal way of life suited to small animals. But with the extinction of the dinosaurs and the break up of Gondwana, mammals survived and flourished.

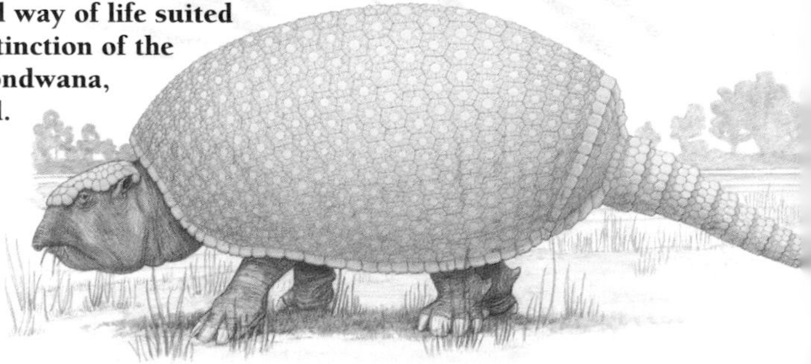

GLYPTODON
The South American mammal fauna of the Tertiary developed in isolation. One group were the edentates including this armored, car-sized relative of armadillos.

Giant mammals North America, Europe and Asia had broken free of Gondwana by 145 mya, but remained connected to one another. The mammals that evolved there were the first of the large, predatory animals, the creodonts. Very like modern carnivores, they had specialized shearing molar teeth and sharp-clawed limbs. Other mammal groups evolved into large herbivores with various patterns of flattened, grinding molars, and elongated legs with little hoofs.

Land bridges Australia, Antarctica and South America remained connected until relatively recently (about 55 mya), but then became completely isolated island continents, on which quite different groups of mammals evolved. In South America, huge edentates evolved, like the armored *Glyptodon* and giant ground sloth *Megatherium*. In Australia, giant kangaroos and wombats occupied the niches taken by placental animals in other parts of the world.

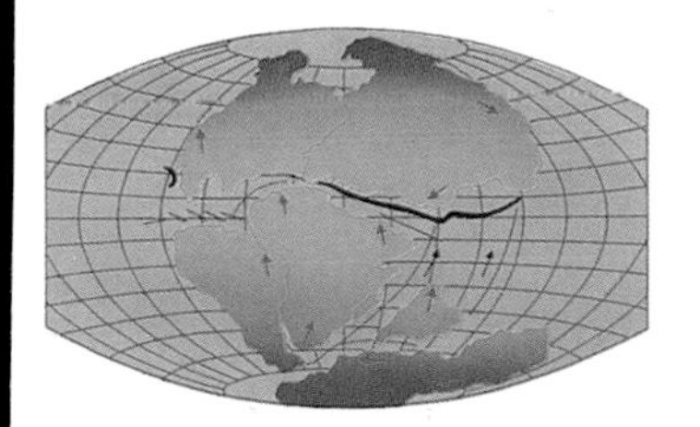

GONDWANA
Before 245 mya, the Earth's continents formed one supercontinent called Pangea. At that time, it began to break up: Laurasia to the north and Gondwana in the south. Continental drift and different land connections over time help explain the distribution of like groups of animals.

Tertiary subera Several major events occurred from around 65 mya. A tenuous connection remained between South America and Antarctica in Gondwana that would soon break apart. New Zealand and Madagascar had already pushed off; southern Africa and Antarctica were also widely separated. The dinosaurs were extinct and a series of climatic changes caused several other extinctions. Mammal evolution was greatly affected by these events, so that within 5 million or so years several important new kinds of mammals evolved. By about 50 mya, the true Carnivora and the earliest of the horse family had appeared, as had the first whales, and other groups later to be highly successful, such as bats, rodents, artiodactyls, elephants and lemurs.

ARSINOITHERIUM
This Egyptian mammal is depicted defending her young against a pack of *Hyaenodon*, 40 mya. *Arsinoitherium* was a browsing herbivore that grew to nearly 13 feet (4 m) long and had a pair of massive horns. She probably derived from the same common ancestor as the elephant.

THE AGE OF MAMMALS

Around 40 mya, the climate deteriorated, causing the extinction of about one-third of the mammal families. The main sufferers were the more primitive kinds, and so the fauna took on a modern appearance. This was followed by a long period of favorable conditions during the Miocene, known as the "Age of Mammals."

Grasslands One of the most significant developments in the Miocene was the spread of grasses, forming great plains for thriving herds of grazing animals. Horses, deer, antelope radiated. The apes also evolved from more primitive primates during this time, and later, about 4 mya, the first members of *Australopithecus* appeared.

Loss of the megafauna The Age of Mammals ended with the disappearance of the megafauna across the world all at about the same time, 10,000 years ago. It is a matter of controversy whether this was due to climatic change, or a result of the rapid spread of humans throughout the world. However, each time humans arrived in a new land, a wave of extinctions appeared to follow: in North America, Europe, Africa, Latin America and Australia.

INDRICOTHERIUM
The largest land mammal ever known was a hornless form of rhinoceros that lived during the Miocene. It attained a height of 18 feet (5.5 m) at the shoulder and weighed perhaps over 40,000 pounds (20 tonnes).

HUMAN COLONIZATION OF THE GLOBE

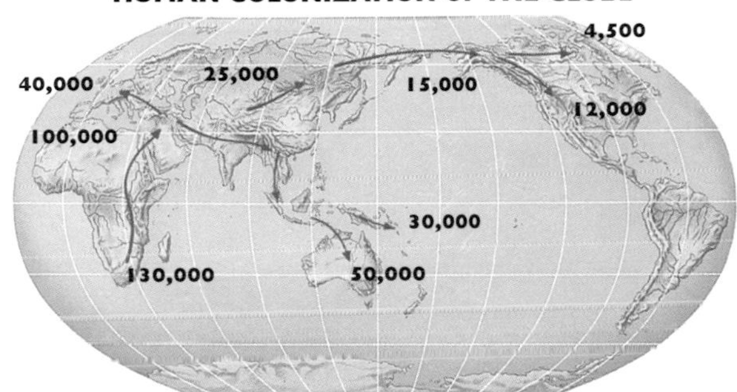

***130,000** Emergence of *Homo sapiens* in south-eastern Africa
100,000 Modern humans appear in south-western Asia
At least 50,000 Arrival of humans in Australia
40,000 First modern humans in Europe
30,000 Seafaring groups colonize the islands east of New Guinea
25,000 Expansion into Siberia
15,000 Humans cross the Bering Strait
12,000 First cultural complex in North America
4,500 First humans in the Arctic
(*Dates are years ago)

LUCY
Lucy is a two-thirds-complete fossil skeleton found in Ethiopia and dating to 3.2 mya. She is only about 3 feet (1 m) tall, with long arms and short legs, and an upright, but not fully human gait. There is no agreement on whether she should be included in the species *Australopithecus afarensis*.

FOSSIL RECORD
Homo sp. The earliest specimen of *Homo* is dated at 2.5 mya from Olduvai in Tanzania. Several of the earliest members of the genus have been found, but only in Africa.
Homo erectus This is the best known species of the so-called "erectines." Their fossils have been found in east Turkana, in Kenya, dating back some 1.6 to 1 million years. The earliest traces of *Homo erectus* outside Africa are found in Java, dated 1 million to 100,000 years ago.
In China, these fossils date from 800,000 to 230,000 years ago; and in Germany fossil specimens are about 300,000 years old. The African, European, Chinese and Javanese fossils are all considered to be subspecies of *Homo erectus*.
Homo sapiens The earliest representatives of our own species—some 130,000 years old—come from Tanzania. *Homo sapiens* fossils have also been found in Israel, dated between 90 and 80,000 years ago.

THE WORLD OF MAMMALS

MAMMAL ADAPTATIONS

Mammals have evolved to successfully occupy almost every habitat on Earth. Much of the great diversity in their external forms reflects adaptations to rainforests, grasslands, tundra, deserts and seas. Mammals of the forests may be excellent climbers with opposable digits and prehensile tails. Grassland dwellers that rely on speed for survival have long, hoofed legs; others may be burrowers. Adaptations for survival in the Arctic tundra include thick, shaggy coats, the ability to turn white in winter, and claws or antlers to expose vegetation. Most desert mammals are nocturnal or have water-retaining metabolisms. Cetaceans have dispensed with hindlimbs and have pectoral fins and streamlined bodies.

COMPARING MAMMALS

Many unrelated animals have evolved remarkably similar social, locomotor, feeding or behavioral adaptations in response to their habitats and ecological roles. This phenomenon is termed convergent evolution. Convergence suggests that there are optimal evolutionary answers to the challenges of adapting to specific habitat types.

Because of their widespread distribution and adaptability, mammals show many excellent examples of convergent evolution.

Convergence of desert rodents

Seed-eating desert rodents exhibit remarkable degrees of convergence in morphology and ecology. Take, for example, the kangaroo-rats of North America, gerbils and jirds in Asia and Africa, jerboas in Asia and Africa and Australian hopping mice. All have pale, sand-colored coats, enlarged hindlegs for rapid movement and long tails.

KOALA AND SLOTH

Many Australian marsupials occupy ecological niches that placental mammals occupy elsewhere in the world. Koalas and sloths are both long-limbed, almost tailless, tree-dwelling leaf-eaters. Neither make nests nor use dens and they carry their young until well developed. They have woolly fur and long grasping claws. They conserve energy by resting in the canopy for up to 20 hours a day.

Ecology of desert rodents

Living in arid environments poses problems of predator avoidance and water retention. Desert rodents have acute hearing, confine their activity to the night, have excellent vision, and sleep in deep, humid burrows during the day. They need little or no water, obtaining sufficient quantities from their food and conserving this by excreting small amounts of highly concentrated urine.

AYE-AYE AND STRIPED POSSUM

This Australian marsupial possum and the Madagascan aye-aye, a primate, have evolved the same highly specialized fourth digit. They feed predominantly on wood-boring insect larvae, exposing the grubs with their protrubing lower incisors, then extracting them from their holes with the claw of this long finger.

PANGOLIN AND SHORT-BEAKED ECHIDNA

Pangolins are unique, scaled mammals from Africa and Asia; the short-beaked echidna is an Australian monotreme—but both are superbly adapted for insect-eating. They have short, powerful limbs for digging into termite mounds and ant nests, and for scooping out their underground burrows. They both have a long, sticky tongue and no teeth. The scales and spines are a formidable defense system when the animals curl into a ball to protect their softer underparts.

NOSES

The sense of smell is highly developed in most mammals, although less important in many primates and reduced or absent in cetaceans. Primates have a highly developed sense of vision; for cetaceans (and many bats) hearing is their most important sense, used in echolocation.

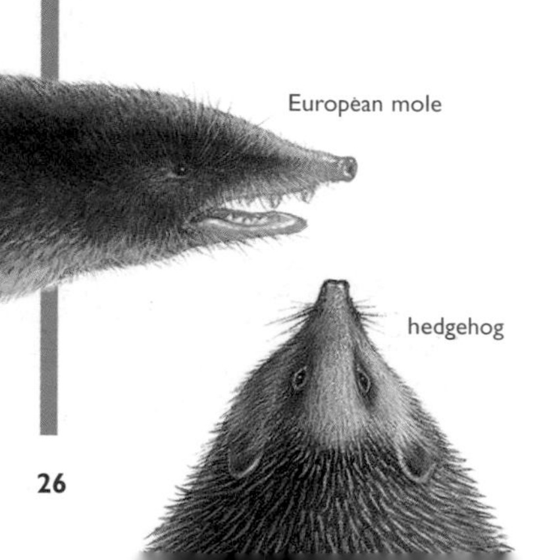

desman

European mole

hedgehog

Probing The Pyrenean desman is a mole, but looks somewhat like a shrew. It hunts its prey underwater, and probes beneath rocks for insects with its long, flexible snout.

Digging A European mole hunts for worms and insects underground. It relies on its sensitive nose to smell and feel out its prey.

Defense The nocturnal Algerian hedgehog is protected by its spiny coat. It has a short, pointed snout with sensitive bristles, and eats a varied diet ranging from insects to mushrooms.

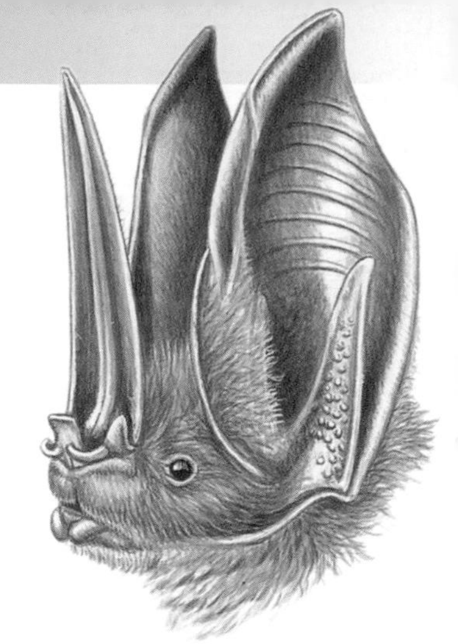

SWORD-NOSED BAT
The New World leaf-nosed bats vary widely in diet, habits and appearance, but generally the muzzle has a spear-shaped noseleaf. These structures aid in directing ultrasonic cries produced through the nostrils. These calls enable the bat to determine the distance and direction of a sound-reflecting object.

IMPRESSIVE PROBOSCIS

The male mandrill has an incredible face. The massive muzzle has bony swellings along the side of the nose. The nose itself is red with blue on the sides; the chin and sides of the face have orange hair; and the lips have red bands. They are forest-dwelling, African primates that eat fruit and leaves, with some insects. In females and young, the colors are more muted.

Insect-eaters Insectivores are animals with long, narrow, often elaborate snouts. Many of the mammal orders contain species that share this specialized diet and adaptation. Over 400 species of small insect-eaters comprise the order Insectivora, including moles, solenodons, hedgehogs and shrews. The order Xenarthra includes the anteaters with their proboscis-like snout. The aardvark, order Tubulidentata, has a long nose for sniffing out termites and resembles anteaters due to convergence, rather than sharing of characteristics.

PROBOSCIS MONKEY

The Bornean proboscis monkey is named for its obvious identifying feature: a large, pendulous nose that hangs below the level of the mouth. Its habitat is coastal mangrove forest and it is a specialized leaf-eater, but the function of the bizarre nose is not known.

EYES

The mammalian eye is well developed and similar in structure and function to the reptilian eye. Color vision has arisen independently in several mammals, including primates and some rodents. Binocular vision, which permits efficient estimation of distance, is particularly well developed in primates.

Binocular vision This is the type of vision where the image of an object in sight falls on the retinas of both eyes at the same time. This is important for accurately judging distance. It gives a three-dimensional effect because the slightly different positions of the two eyes enable the object to be viewed from slightly different angles. The position of the eyes at the front of the head, as opposed to a lateral arrangement, is important for the functioning of binocular vision. Primates, fruit bats, flying lemurs and cats are mammals with binocular vision.

JUDGING DISTANCE
The frontal positioning of the eyes is extremely important for predators, like this lion hunting a gazelle. It permits binocular, or stereoscopic, vision to accurately judge distance between itself and the prey.

Eye shine Many nocturnal mammals and almost all Carnivora have a reflective layer at the back of the eye that assists in night vision. The layer is called the tapetum lucidum, and it reflects weak incoming light back onto the retina, giving the sensory cells a "second chance" to respond. Eye shine is the reflection that comes back when a bright light is shone into one of these animals' eyes, caused by the presence of the tapetum lucidum. The eyes are therefore more sensitive for night vision, but visual effectiveness is lessened because the reflection is not perfectly clear and the image is further blurred.

TARSIER
The tarsier is a small, nocturnal, carnivorous primate from South-East Asia. The eyes are truly enormous, and in the most specialized species, the orbits are so flared that the skull is broader than it is long.

DOUROUCOULI
The night monkey, from the forests of South and Central America, is the only nocturnal monkey. Like the tarsier, it has no tapetum lucidum to aid in nocturnal vision. Both species compensate by possessing the largest eyes of all the primates.

EARS

Hearing is important to most mammals and a number of specializations have evolved, particularly the amplifying function of the middle ear. The external ear collects sound and concentrates it onto the opening to the middle ear. Many mammals can hear sounds of a very high pitch, and this has been exploited by bats, for example, for echolocation.

Echolocation This is the way animals use sound and hearing to detect the position of an object. Echolocation is an adaptation used by bats and cetaceans that have poor vision. The Microchiroptera, small, insect-eating bats, project ultrasonic calls through the mouth or nostrils. The calls vary in intensity and frequency in accordance with the bat's own behavior in moving through its habitat. Objects reflect the sounds back to the bats that are then able to identify the distance and direction of the object.

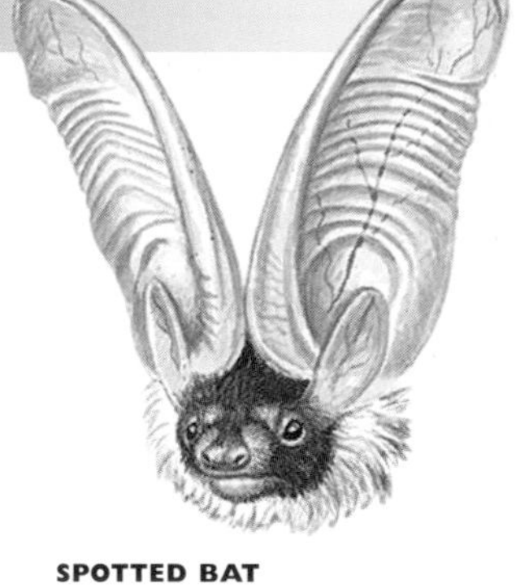

SPOTTED BAT
This member of the diverse family Vespertilionidae has exceptionally long ears, which aid in collecting reflected sound in the complex process of echolocation. The tragus is a small, projecting flap just inside the ear that concentrates the sound and directs it to the opening.

CARACAL
The small African caracal's ear tufts are the longest of any cat species. It has been suggested that the tufts serve as antennas to aid hearing, but they are more probably used to accentuate facial expressions. Caracals "ear flick" at one another as a threat.

No ears? Ears may appear to be absent in species such as moles, which are adapted for a subterranean life. This is due, however, to a lack of external ear flaps and the covering of thick fur on the ear opening: in fact, their hearing is acute. Cetaceans, also, do not have external ears, but have ear openings behind the eyes. They too have excellent hearing over enormous distances. The eyes and ears of the platypus are sealed by a fold of skin controlled by strong muscles when it dives.

Big ears The enormous ears of the African elephant or the black-tailed jack rabbit, in addition to their function in hearing, act as huge radiators to help control the body temperature (and for elephants make useful fans). When the blood vessels of the ears are engorged with blood, the animal loses heat; when constricted, body heat is conserved. An elephant's ears are also used as signals of emotional state.

SOUND FACTS

Sound travels five times faster in water than in air.

In bottlenose dolphins, echolocation is effective to at least 2,500 feet (762 meters).

Sound produces a three-dimensional picture of an object's texture and internal structure.

Sperm whales have many tons of fat deposits in their brain case and lower jaw that act to focus sound produced in echolocation, and receive and transmit the response to the whale's ear and brain.

ELEPHANT EAR COMPARISON

The most obvious difference between the African and Asian elephant species is their ear size: those of the African are larger. Anatomical differences between the subspecies are less obvious and mainly due to adaptations to different habitats. The forest African elephant occupies more areas with canopy forest than the bush subspecies. The bush elephant has triangular ears that extend below the line of the neck; the forest elephant's ears are rounder and do not extend below the neckline.

forest African elephant

bush African elephant

HORNS AND ANTLERS

Tusks, horns and antlers are very different structures, but they are used for similar purposes by mammals. Males use these weapons in disputes over territory or mates, or to assert their dominance in a group. They are symbols of age, strength and status and can actually help in avoiding physical combat.

SKIN AND BONE

Tusks, horns and antlers are made of different substances. Walrus and elephant tusks are actually overgrown teeth. Bongo, cow and gazelle horns are permanent bony structures covered with a hard skin called keratin. Rhinoceros horns are also made of keratin and like all skin, will grow back if worn or cut off. The antlers of deer, elk and moose are covered with a soft skin called velvet, and are grown and shed annually. They are made of bone and grow from protrusions on the head called burrs.

Rhinoceros Rhinos bear horns on their snouts. Unlike the horns of antelope, those of rhino lack a bony core. They consist of a mass of hollow filaments that adhere together and are attached fairly loosely to a roughened area of the skull. These animals are being pushed toward extinction due to the high value of the horns in some Asian countries, as medicine. The horn is made of the same substance as cattle and antelope horns, hooves and human fingernails: keratin. Rhino horn will regrow if removed, making the prospective loss of the species even more devastatingly pointless.

Tusks Walrus tusks are upper canine teeth and elephant tusks are upper incisors. Like all mammalian teeth, they are composed of dentine, with pulp cavities containing blood vessels and nerves.

THE UNICORN LEGEND

The unicorn is a mythical creature with the body of a horse and a single, long horn on its head. The legend probably originated from exaggerated descriptions of rhinoceroses. In the Middle Ages, people believed that unicorns had magical abilities. They thought that if they drank from a cup made of "unicorn" horn (possibly acquired from a narwhal tooth or rhinoceros horn) they would be protected against poisoning..

STATUS SYMBOLS

Each year, a moose loses its antlers and grows more. The antlers become bigger and heavier as the moose gets older. Antlers are signs of the male's status and signal dominance when in competition for food and serve to attract females. Rival male moose will confront each other first, and if neither retreats will lock antlers in a battle of strength and endurance.

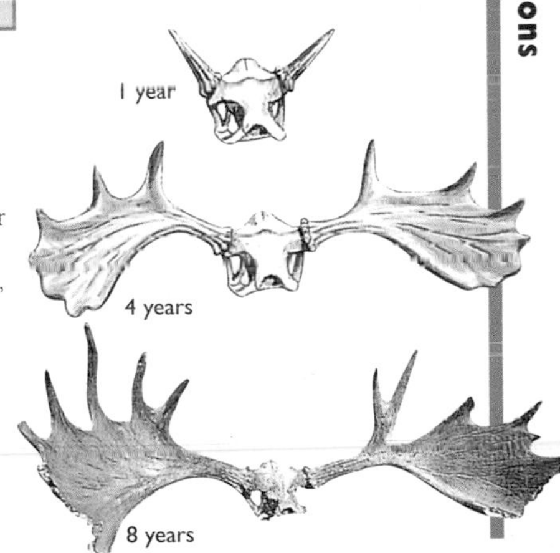

Defense strategy Horns and antlers represent a defense strategy to deter attackers without risk of wounding either party. (Blood attracts predators, no matter whose blood it is.) This is particularly important when members of the same species are competing for mates, food or territory.

Weaponry Short and sharp horns, or outgrowths on the head that enable it to be used as a club, are weapons which provoke pain but will not stick in an opponent's body. They serve to make the intruder retreat and discourage retaliation.

Sparring Horns and antlers that have adapted into twisted, rugose or branched "baskets," that readily catch and bind the opponent's head, permit head wrestling. These are harmless weapons that encourage sparring. Simple grappling horns defend mating territories, but in species with harem defense of females, huge antlers or horns may evolve, as in reindeer, the extinct Irish elk and the water buffalo.

CLAWS

Claws and talons are stronger, sharper versions of human nails. They are used for grabbing prey, climbing, burrowing, digging, hooking or slashing a rival. Many carnivores have claws that are unique in being retractile, saving them from wear, and allowing them to climb and kill with great efficiency.

Uses of claws Polar bears seize seals in their clawed forefeet, while grizzly bears dig tubers with theirs. Sturdy claws are also useful for climbing trees. Koalas and sloths share the feature of strong, grasping claws. Digging and burrowing for food or shelter is aided by specialized claws, like those of wombats. They are also used in defense. The silky anteater has an immense, hook-like claw on each forefoot and when threatened, rears up on its hindfeet to its full size, wielding and slashing at its would-be attacker.

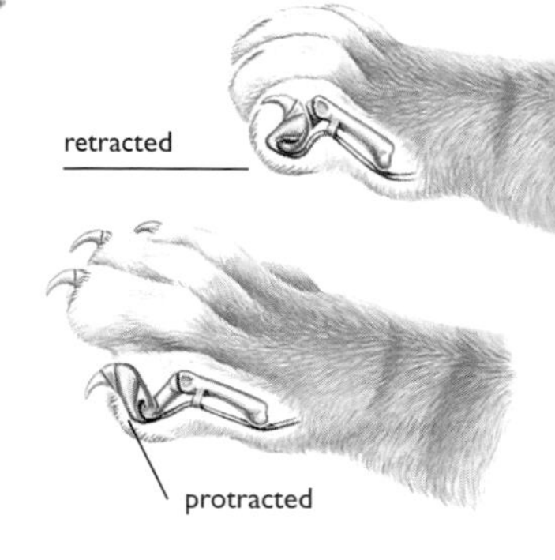

JACKKNIFE
A cat's prey must be well controlled to enable the neck bite that is typical of their predation. This control is provided by sharp claws. Because the claws are usually retracted inside fleshy sheaths in the paw, they are not subject to constant wear. Claws are in the normal, relaxed state when retracted. Muscular contractions cause a spring ligament to stretch and protrude the claws beyond the sheath. The claw mechanism works just like a switchblade.

ARCTIC ADAPTATION

As an aid to digging through snow, ice and frozen ground for food in the Arctic, the collared lemming has a seasonal adaptation. The claws on the third and fourth digits of the forefeet become greatly enlarged each winter.

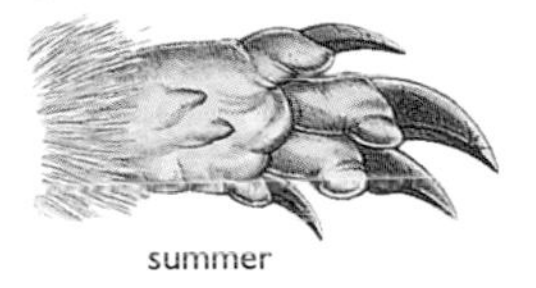

summer

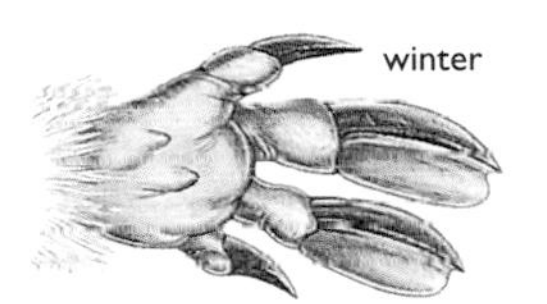

winter

BEAR PAWS

The claws of bears are not retractile, and vary in size and shape according to their use. The black bear (right) has curved, hook-like claws for climbing. The giant panda (far right) has an enlarged pad, the shape of an upside-down "L" that enables it to grasp bamboo.

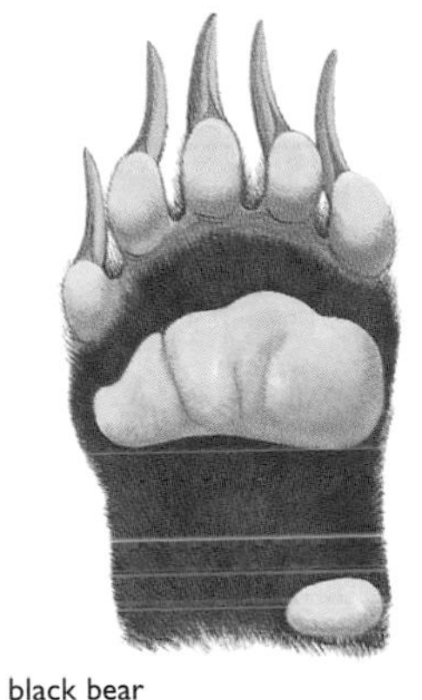

black bear forefoot

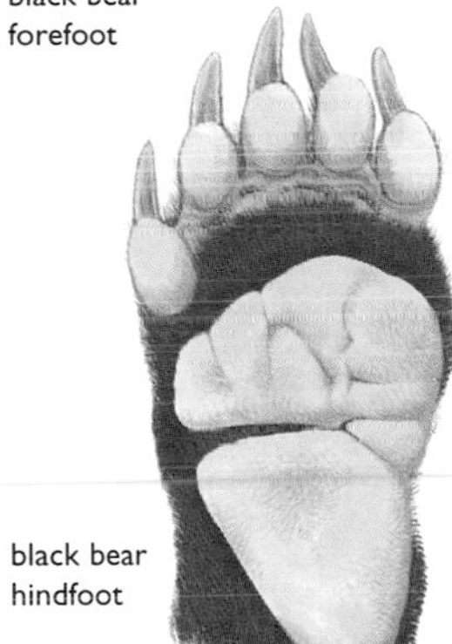

black bear hindfoot

giant panda forefoot

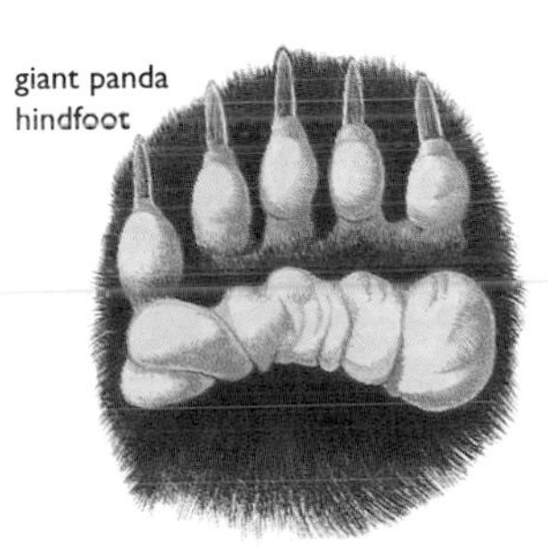

giant panda hindfoot

TAILS

The mammalian tail is a continuation of the spine, but varies widely in function and form. The long tails of forest dwellers, like mice and rats, act as counterbalances as the animals move along branches. More specialized arboreal species have prehensile tails that serve as a fifth hand. The tail may also be used for communication.

Fifth limb Prehensility is the adaptation for grasping, seizing or holding. There are degrees of prehensility in mammal tails, from the firm hold on the branch by a howler monkey or opossum, to the only slightly prehensile tail of a musky rat-kangaroo, which uses its tail to push together nesting material.

Other uses The bushy tails of squirrels, when recurved over the body, provide shade. An anteater lies on its side and uses its tail as a blanket. Aquatic rodents, such as the beaver, or the monotreme platypus, have horizontally flattened tails for swimming, which produce "lift" or "negative bouyancy." Rapid runners and jumpers commonly have a long, streaming tail for a counterbalance, especially the fastest of all land animals, the cheetah.

WOOLLY SPIDER MONKEY
The rare muriqui inhabits the Brazilian rainforest. It moves around the forest canopy in groups, eating fruit, seeds and leaves. Muscular prehensile tails have evolved independently in diverse mammal groups in response to the challenges of living in the forest.

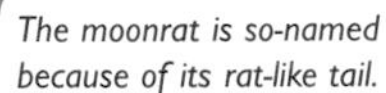

The moonrat is so-named because of its rat-like tail.

The Pyrenean desman is a type of swimming mole.

ROO TAILS

When moving slowly, the kangaroo tail is pressed to the ground while the hindlimbs are raised, forming a tripod with the forelimbs. The hindlimbs are then swung forward, always together, because a typical kangaroo is unable to walk. In this way the tail acts as a fifth limb in the slow gait, but has a balancing function in the fast hopping gait.

SCALY TAILS

The tails of some mammals are covered with scales rather than hair. The animals featured use this adaptation in their variable aquatic lifestyles. Moonrats prefer wet, swampy habitats and often enter the water to hunt. Desmans live in mountain streams, using their broad tails for steering. Desmans and water opossums both dive for bottom-dwelling invertebrates.

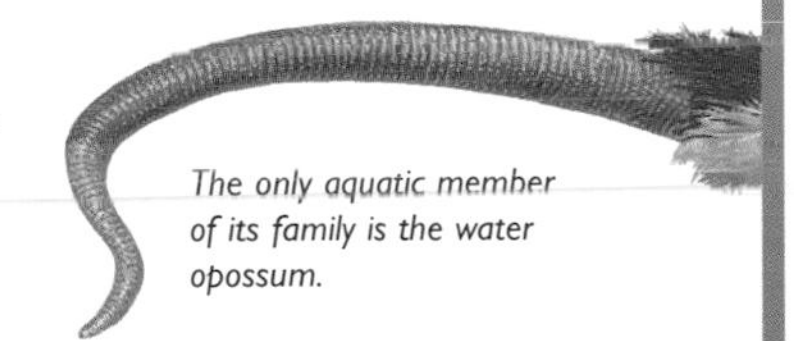

The only aquatic member of its family is the water opossum.

PART TWO

KINDS OF MAMMALS

MAMMALS GREAT AND SMALL

Because mammals are warmblooded, they have been successful in almost every environment. They are one of the smaller classes of animals, but the variety among them is astounding. Their great diversity has evolved for over 200 million years, resulting in body shapes and functions, and behavioral adaptations for their survival in specialized ecological niches.

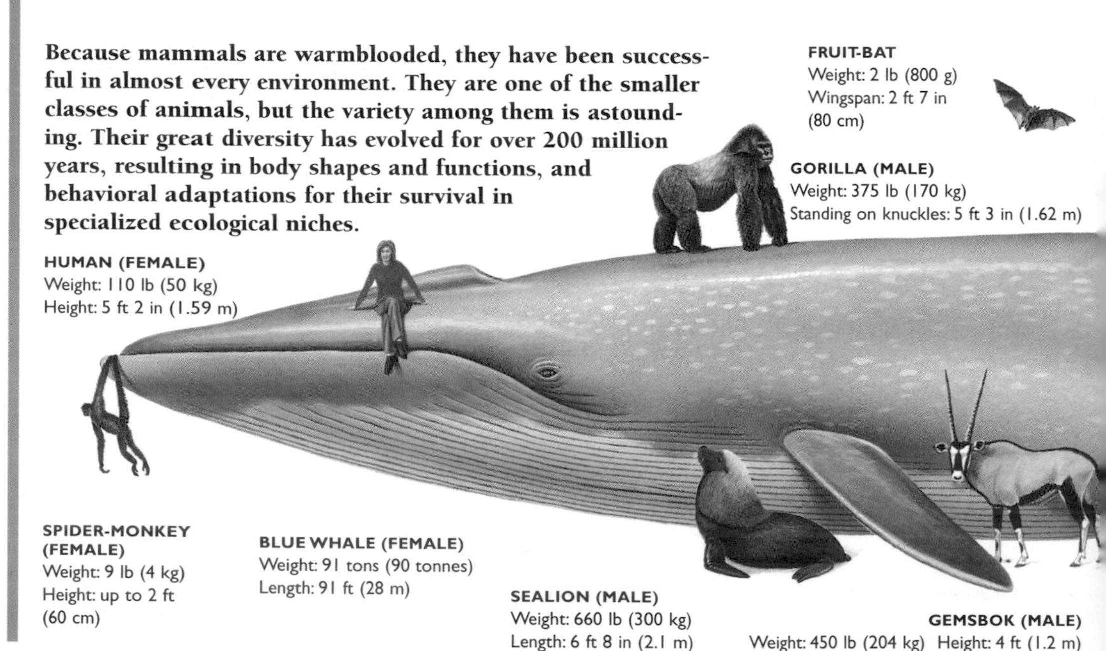

FRUIT-BAT
Weight: 2 lb (800 g)
Wingspan: 2 ft 7 in (80 cm)

GORILLA (MALE)
Weight: 375 lb (170 kg)
Standing on knuckles: 5 ft 3 in (1.62 m)

HUMAN (FEMALE)
Weight: 110 lb (50 kg)
Height: 5 ft 2 in (1.59 m)

SPIDER-MONKEY (FEMALE)
Weight: 9 lb (4 kg)
Height: up to 2 ft (60 cm)

BLUE WHALE (FEMALE)
Weight: 91 tons (90 tonnes)
Length: 91 ft (28 m)

SEALION (MALE)
Weight: 660 lb (300 kg)
Length: 6 ft 8 in (2.1 m)

GEMSBOK (MALE)
Weight: 450 lb (204 kg) Height: 4 ft (1.2 m)

DEPENDENCE ON ENVIRONMENT
There are only about 4,300 mammal species. Habitat destruction, environmental abuse and high-technology killing is causing an "extinction spasm." Experts estimate that we are losing up to 100 species of plants and animals a day. If this alarming trend continues, then virtually all mammals will be endangered by 2008.

GIRAFFE (MALE)
Weight: 2,600 lb (1,180 kg)
Height: 16 ft (4.95 m)

GROUPING ANIMALS
The animals shown here have some very obvious differences, though they all share the common features of the class Mammalia. Their scientific classification reveals much more about the similarities between them than mere body form. Animals that share an order, like gorillas and monkeys, have less in common than members of the same family, like humans and gorillas.

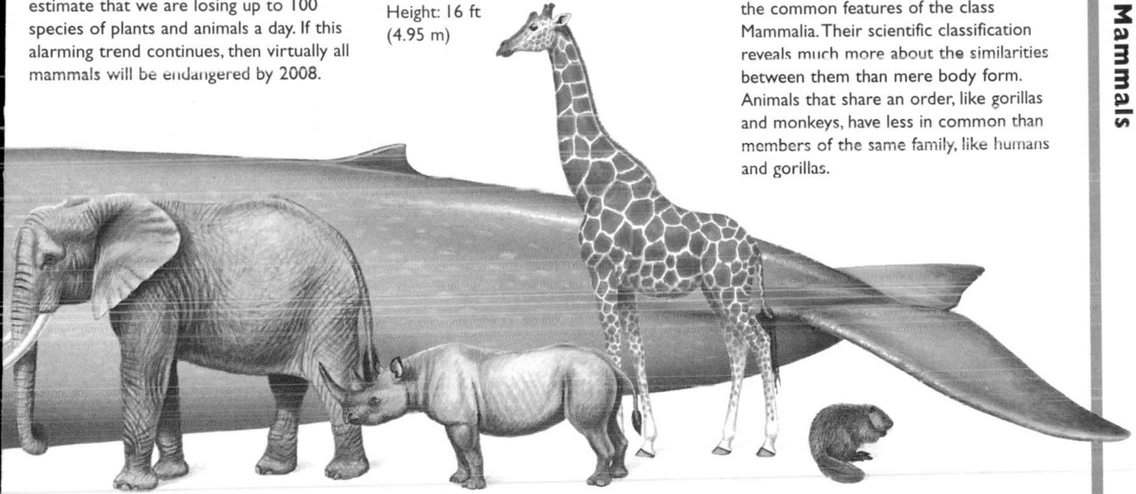

AFRICAN ELEPHANT (MALE)
Weight: 5 tons (5.1 tonnes)
Height: 11 ft (3.35 m)

BLACK RHINOCEROS (MALE)
Weight: 1.25 tons (1.3 tonnes)
Height: 5 ft (1.52 m)

BEAVER
Weight: 66 lb (30 kg)
Length: 3 ft 3 in (1 m)

KINDS OF MAMMALS

MONOTREMES

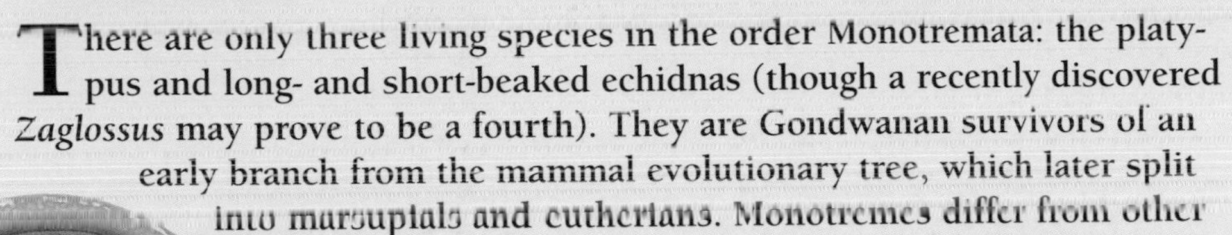

There are only three living species in the order Monotremata: the platypus and long- and short-beaked echidnas (though a recently discovered *Zaglossus* may prove to be a fourth). They are Gondwanan survivors of an early branch from the mammal evolutionary tree, which later split into marsupials and eutherians. Monotremes differ from other mammals in laying eggs and lacking teats, though they secrete milk from mammary glands. The name Monotremata means "one hole" for the single opening through which waste products and female eggs are passed. This feature is not unique to monotremes, but is shared by marsupials.

Short-Beaked Echidna

The short-beaked echidna has been successful as it is the only Australian mammal to feed on ants. Long-beaked echidnas existed in Australia until around 15,000 years ago, possibly dying out due to competition with other species for food resources on the forest floor. Its surviving descendant is rare in New Guinea.

Close relatives There are two species of echidnas.

Characteristics Echidnas have a snout which forms a long beak-like organ with nostrils and mouth at the end. They have a long, protrusible, sticky tongue and no teeth.

CLASSIFICATION

ORDER MONOTREMATA
FAMILY TACHYGLOSSIDAE

TACHYGLOSSUS ACULEATUS
The spiny coat forms an excellent defense. Spines begin to grow at about nine weeks and the young is understandably moved from the pouch to the burrow at this time.

Monotreme distribution

The order is restricted to New Guinea and Australia and fossil evidence points to a Gondwanan origin.

Food The short-beaked echidna eats termites and ants.

Young A pouch develops during the breeding season when one egg is laid. After 10 days it hatches to suckle from milk patches in the pouch.

Habitat Short-beaked species are common, with no particular habitat requirement other than one with a food supply.

Platypus

This ancient, small, amphibious animal is one of the most unusual creatures alive. Platypus teeth from Patagonia are 60 myo and Australian fossils are 130 myo. The living platypus has teeth that develop in the embryo, but are resorbed, showing a physical link with the fossil species.

Close relatives The platypus is monotypic in its family.

Characteristics It has a coat of dense, insulating fur; a pliable, sensitive, bill; and webbed forefeet. It occupies a burrow in river banks and is 15–25 inches (39–63 cm) long.

Food The platypus dives for freshwater invertebrates in two-minute intervals, with eyes and ears closed, locating food with its bill. Above water it chews the food collected in cheek-pouches.

Young After three weeks' gestation, one to three eggs are laid. They are incubated for 10 days between the tail and body in a nesting burrow that may be 66 feet (30 m) in length. There the young remain for three to four months, feeding on milk oozing from patches on the underside of the body.

Habitat The species is common in the streams, rivers and lakes of much of eastern Australia.

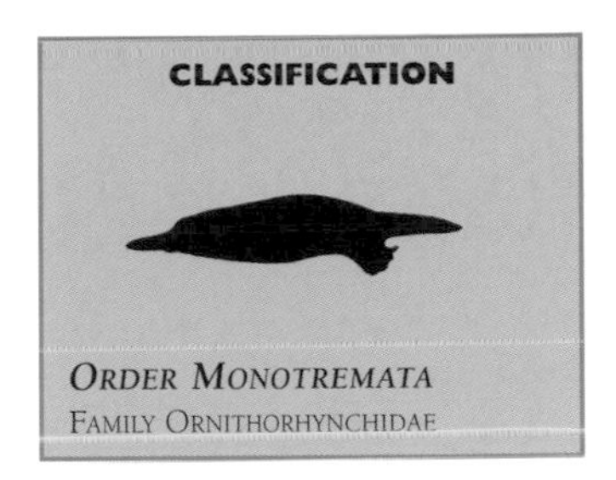

ORNITHORHYNCHUS ANATINUS
Able to maintain a stable body temperature and with a coat of dense fur, the platypus can swim for long periods in near-freezing water. Longer, flattened hairs project through the underfur. These molt throughout the year whereas the underfur remains intact and dry. The platypus spends much time grooming, both in and out of the water.

MONOTREME ADAPTATIONS

When the first platypus specimen was sent to England in 1798, many experts decried it as a fake, made by stitching the bill of a duck and a beaver's tail to the body of an otter. However these adaptations in body form and function have secured it, and its relative the echidna, ecological niches with little competition.

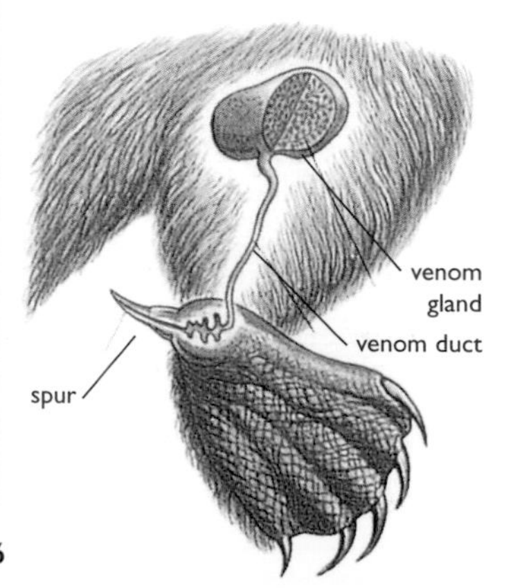

POISONOUS SPURS

Rear ankle spurs are present in all three monotremes. In females, the spur is lost, and although spurs exist in echidnas, they do not use them to inject venom as the male platypus does. Fully grown spurs are 1/2 inch (1.2 cm) long and driven into an object by the rear leg muscles. The wound is painful, but the venom can lead to symptoms ranging from local pain and swelling to paralysis of a limb in humans. The spurs are certainly a deterrent to predators, but their loss in females suggests this is not the primary function, which may be related to establishing breeding territories.

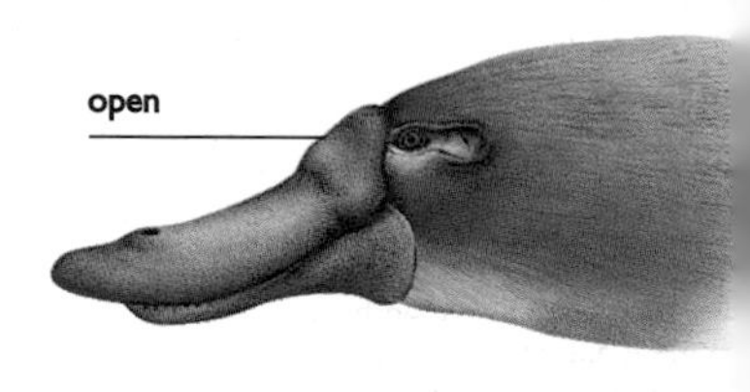

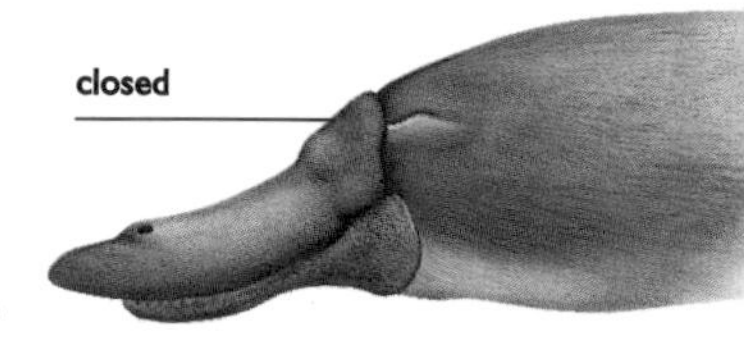

WATERTIGHT

A groove of skin, controlled by strong muscles, encloses the eyes and ears of the platypus when diving for food. Prey is located in often murky waters by the electrosensitive and touch-sensitive, rubbery bill.

LONG STICKY TONGUE

Both species of echidnas feed by means of a protrusible tongue, made sticky by mucous secretions. The short-beaked echidna may consume thousands of ants during a day's feeding, excavating a nest with large claws, probing its snout into small spaces, and extending its tongue into cavities to gain access to the insects. The generic term, *Tachyglossus*, actually means "swift tongue." The long-beaked echidna is adapted for worm eating. Spines on a groove in its tongue hold the prey fast. Both species lack teeth and chew food between a horny pad at the back of the mouth and on the palate.

SURVIVAL SPECIALIST

Echidnas are long-lived: a short-beaked echidna in the Philadelphia zoo lived for 49 years, and in the wild a marked individual was 16 years old. This is partly due to the formidable defense system of a spiny coat and legendary vertical-burrowing ability, that allows them to disappear into the soil in less than a minute. When disturbed on open, hard ground, the echidna curls into a ball almost completely covered with spines. However, the belly remains vulnerable and an experienced dingo may penetrate this spot. It prefers to dig itself into the soil for protection with its spines and limbs stretched out. Fur is present between the spines and in cooler environments the fur may be long enough to cover the spines.

KINDS OF MAMMALS

MARSUPIALS

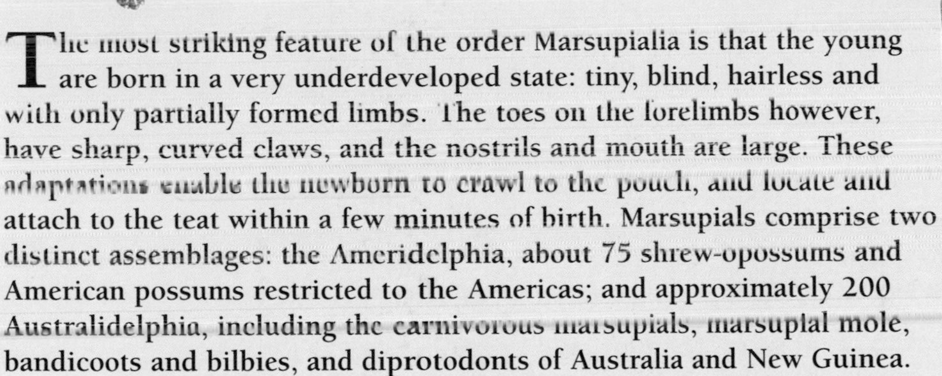

The most striking feature of the order Marsupialia is that the young are born in a very underdeveloped state: tiny, blind, hairless and with only partially formed limbs. The toes on the forelimbs however, have sharp, curved claws, and the nostrils and mouth are large. These adaptations enable the newborn to crawl to the pouch, and locate and attach to the teat within a few minutes of birth. Marsupials comprise two distinct assemblages: the Ameridelphia, about 75 shrew-opossums and American possums restricted to the Americas; and approximately 200 Australidelphia, including the carnivorous marsupials, marsupial mole, bandicoots and bilbies, and diprotodonts of Australia and New Guinea.

Woolly Opossum

The American marsupials comprise three families: the American opossums (Didelphidae), shrew opossums (Caenolestidae) and the single colocolo (Microbiotheriidae). Most are omnivorous and able to climb. The first toe of the hindfoot is opposable to the other four and lacks a claw. The woolly opossums are arboreal.

Close relatives The mouse-, short-tailed, lutrine and other opossums comprise the family, with three species of woolly opossums in the genus.

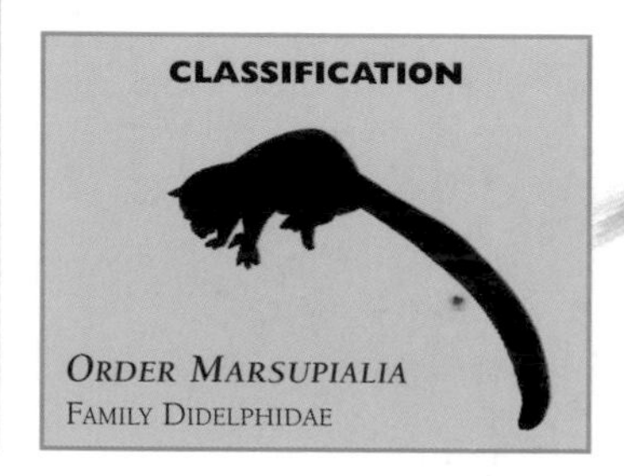

CLASSIFICATION

Order Marsupialia
Family Didelphidae

Food The woolly opossum is a frugivore that eats mostly fruit and nuts.

Characteristics The woolly opossum spends its life in the forest canopy. It has large forward-facing eyes and a prehensile tail.

Habitat Woolly opossums inhabit tropical rainforest in Central and South America

Young The reproductive biology of the Ameridelphians varies from the Australians only in having paired sperms.

Marsupial distribution

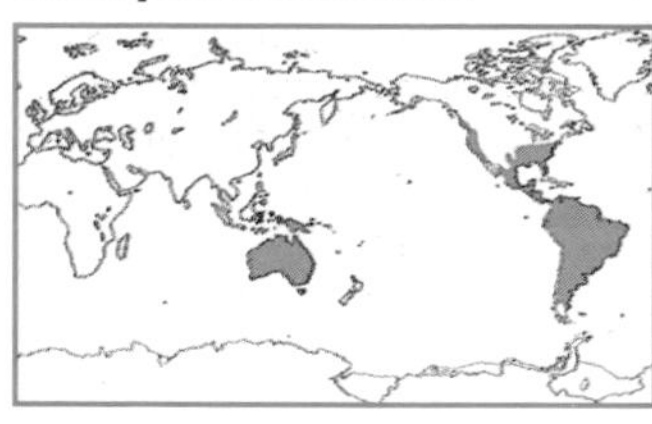

The Ameridelphia are restricted to the Americas, and Australidelphia to Australia, New Guinea and nearby islands.

***CALUROMYS* SP.**
The woolly opossum, genus *Caluromys*, is a medium-sized "true" opossum. It is a tree-dweller with a monkey-like appearance and differs little from the Australian possums, beyond its distribution.

Water Opossum

The water opossum, or yapok, is the only truly amphibious species of American opossum, although most marsupials can swim.

Close relatives The yapok is monotypic in its genus.

Characteristics It has webbed hindfeet and a scaly tail, which is shorter than in the arboreal species and used as a rudder. Its eyes are shut when underwater and prey is located by probing with its sensitive fingers.

Food The water opossum dives for invertebrates, fishes and amphibians on the bottom of streams or lakes.

Young The rear-opening pouch is closed by a strong sphincter muscle and sealed with water-repellent secretions when the animal is swimming.

Habitat This species is distributed around freshwater ponds and streams, from southern Mexico to northern Argentina.

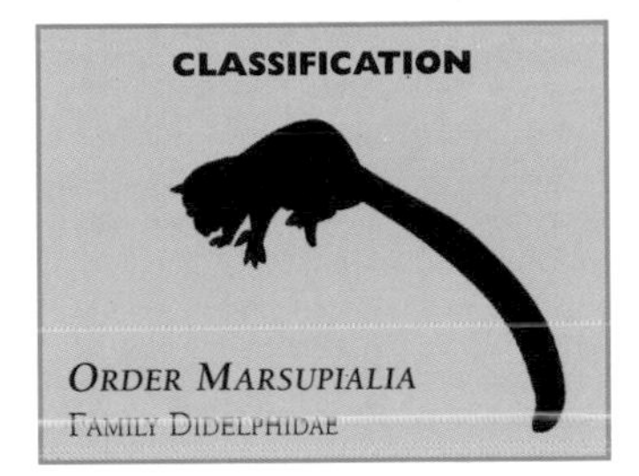

CHIRONECTES MINIMUS
There are approximately 75 species of Ameridelphians, all but one of which are referred to as opossums. Most can swim and some enter the water to swim after prey but only the yapok is adapted to an aquatic lifestyle.

Spotted-tailed Quoll

Quolls are classified as carnivorous Australian marsupials in a family of rather unspecialized predators. The spotted-tailed quoll is the largest of the mainland carnivores. The group includes the numbat (Myrmecobiidae) and the extinct Tasmanian tiger (Thylacinidae).

Close relatives
The dasyurids are the quolls, antechinuses, Tasmanian devil, kowari and mulgara.

Characteristics There is little variety in body shape among the family, though the spotted-tailed quoll is distinguished by its rich brown hair with white spots. It has particularly large, saber-like canine teeth. It is partly arboreal and has ridges on the pads of its feet, but lacks a prehensile tail.

Food It is an efficient predator, hunting prey from insects to small wallabies. Birds and carrion also form part of the diet.

Young An average of five young are housed for seven weeks in the permanent, deep pouch with six teats.

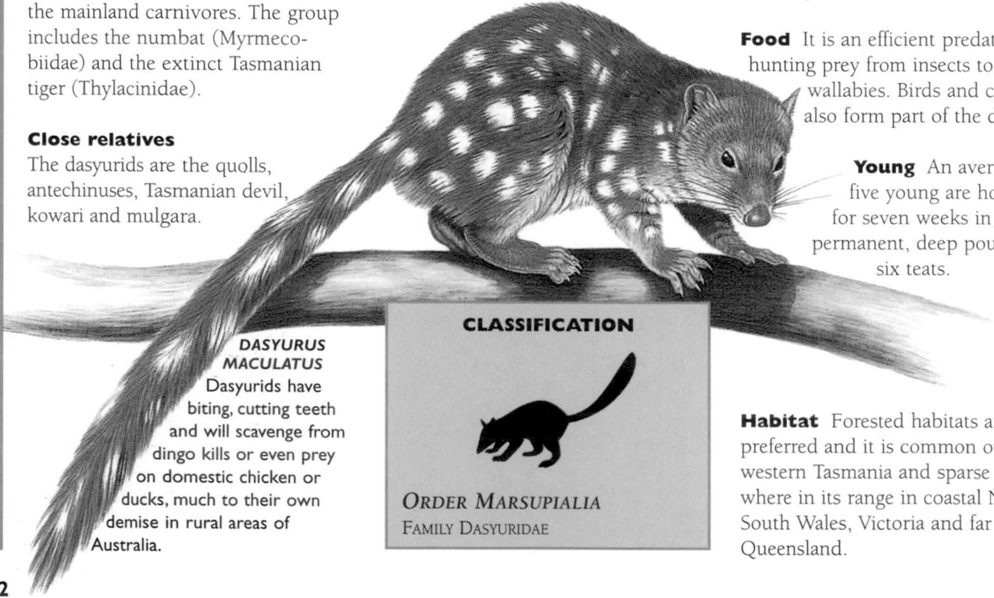

DASYURUS MACULATUS
Dasyurids have biting, cutting teeth and will scavenge from dingo kills or even prey on domestic chicken or ducks, much to their own demise in rural areas of Australia.

CLASSIFICATION

Order Marsupialia
Family Dasyuridae

Habitat Forested habitats are preferred and it is common only in western Tasmania and sparse elsewhere in its range in coastal New South Wales, Victoria and far North Queensland.

Tasmanian Devil

Another of the unspecialized, carnivorous marsupials, the "devil" resembles a stocky dog, but one with a pouch that can climb trees.

SARCOPHILUS HARRISII

Despite its fearsome name and appearance, the devil is a shy creature, and its disappearance on the Australian mainland is partly due to predation by dingos.

CLASSIFICATION

ORDER MARSUPIALIA

FAMILY DASYURIDAE

Close relatives The dasyurids include the antechinuses, Tasmanian devil, quolls, kowari and mulgara.

Habitat The Australian island of Tasmania.

Characteristics The devil is black all over with white marks on the chest and powerful jaws.

Food It usually forages alone, eating birds, insects, beached fish, possums, wallabies and wombats.

Young Up to four young are carried in a permanent, shallow pouch with the corresponding number of teats. After four months they are fully furred and are moved to the den.

NUMBAT

The numbat is Australia's only diurnal marsupial. It has a striking coat, but is similar in body form to the other marsupial carnivores.

Close relatives The numbat is monotypic in its family.

Characteristics A long, sticky tongue flicks out to gather food. It does not have the specialized digging forelimbs of other insect-eaters, but does dig burrows for sleeping nests. The tail hairs are long, giving a brush-tailed appearance.

Young Birth is 14 days after conception and young are suckled on the four teats for six to seven months. They are then placed in the burrow, and weaned by 10 months old.

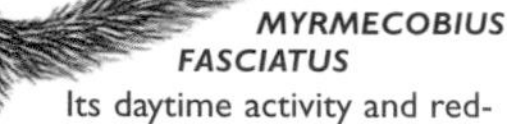

MYRMECOBIUS FASCIATUS

Its daytime activity and red-brown coat with a darker rump and white stripes set the numbat apart from its less colorful and generally more secretive marsupial relatives.

CLASSIFICATION

ORDER MARSUPIALIA
FAMILY MYRMECOBIIDAE

Food The numbat is primarily a termite eater, but some ants are taken incidentally. It is not able to dig into termite nests, but exposes runways just below the surface.

Habitat Eucalypt forest provides hollow logs for shelter and termites for food. The numbat digs burrows or builds nests in logs to sleep in. It is now rare in a few isolated populations in south-western Australia and is considered seriously endangered.

BILBY

Bilbies and bandicoots form a subgroup of marsupials known as peramelemorphs. They are all terrestrial with long, pointed heads and compact bodies. In common with macropods, the second and third toe of the hind foot are fused together.

Close relatives The Australian bandicoots and the single surviving bilby form the family.

MACROTIS LAGOTIS
The bilby is a powerful burrower and may dig holes nearly 6 feet (1.8 m) deep and 10 feet (3 m) wide.

Habitat The bilby occupies a range of desert habitats in central Australia, but the population is thought to be contracting into smaller units. A second, smaller, white species, the lesser bilby, was last reported alive by Aborigines from the central deserts in the 1960s. Bilbies are at risk from predation by foxes and cats, and in competition for food with rabbits and cattle.

Characteristics The bilby has long ears and long, blue-gray, silky fur. It has strong forelimbs with powerful claws for digging and burrowing. Its long hindlimbs are used for leaping, though it usually moves in a galloping motion. It is strictly nocturnal, spending the daylight hours in a deep burrow.

Food The bilby eats small mammals and lizards, as well as insects and their larvae, seeds, bulbs, fruit and fungi.

Young The pouch has eight teats but is usually occupied by two young for 80 days.

CLASSIFICATION

ORDER MARSUPIALIA
FAMILY PERAMELIDAE

Koala

The koala is the only living member of its family. Together with the wombats, possums and macropods (kangaroos and their kin), they make up more than half the Australian marsupial fauna. This group is collectively called the Diprotodontia.

Close relatives Koalas are most closely related to wombats, though not sufficiently so to share a family.

Characteristics The koala is a tree-living leaf-eater. It is long-limbed and, in common with the wombat, is almost tailless. It has woolly fur that is short and pale gray in northern animals, and longer and gray-brown in the cooler south. The larger animals may weigh over 24 pounds (12 kg).

PHASCOLARCTOS CINEREUS
The young koala continues to associate with the mother, travelling around on her back until fully weaned at about 12 months of age and quite large.

KOALA CLAWS
The koala can quickly bound up a tree, grasping it with its sharp forepaw claws. It moves along branches using the grip of its opposable digits on the forepaw and foot. The second and third toes of the hindfoot are fused, and used as a comb.

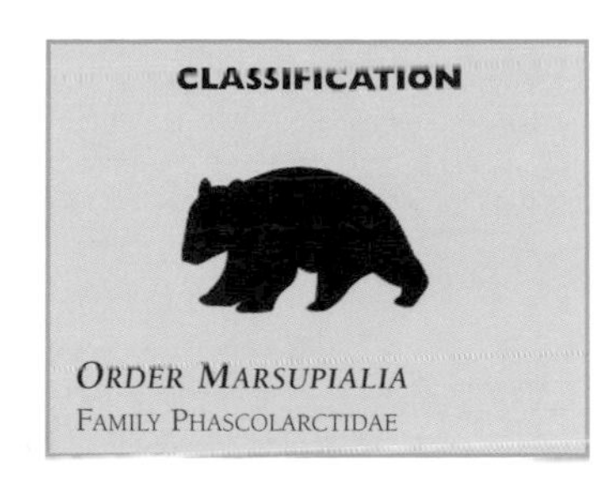

ORDER MARSUPIALIA
FAMILY PHASCOLARCTIDAE

ADAPTED TO LAZE
A koala rests for about 20 hours a day in a low tree fork. They climb into the canopy around dusk to feed and descend to the ground during the night, to change to a preferred feeding tree.

Food The koala is directly associated with eucalypt forest, feeding almost exclusively on the leaves of these trees. There are clear preferences among regional koala populations as to the preferred species of eucalypt. It has developed unique behavioral and physical adaptations to be able to exist on the barely digestible foliage. Koalas have low energy requirements, powerful jaws, sharp teeth and a specialized digestive system. The caecum, in the hindgut is very large and is used for fermenting food products.

Young Females produce one young a year for up to 16 years, giving birth after a gestation of 35 days. The young attaches to one of the two teats in the pouch, and stays there for about six months. It is first seen when it pokes its head out of the pouch and leaves the pouch permanently one month later. It feeds on both milk and eucalyptus leaves until fully weaned at 12 months old.

Habitat The koala occupies a large but fragmented area, the length of eastern Australia. Different populations prefer different eucalypt species and larger populations occur with eucalypt species that grow in richer soils. The koala is widespread and has no natural predators, but some groups are at risk when dense populations live in reducing habitat.

WOMBAT

The three species of wombats—the common wombat and two hairy-nosed wombats—are members of the diprotodont group. Like koalas, they have a vestigial tail, two teats in the pouch, cheekpouches and a special stomach gland. Like kangaroos they are grazers. They are unique in the group in their adaptation for building elaborate burrow systems.

Close relatives Wombats are most closely related to koalas.

Food Wombats favor green shoots.

Characteristics Wombats are stocky creatures with stout heads, short, broad feet and powerful limbs. They are mainly nocturnal, and weigh an average of 70 pounds (30 kg). Ear length and nose shape vary between species.

VOMBATUS URSINUS The common wombat inhabits forests and grazes at night in clearings.

Young The young spend six months in the pouch and stay close to the mother for another 11 months.

Habitat In their limited ranges two species are common, but the northern hairy-nosed wombat is rare and endangered.

LASIORHINUS SP.
The hairy-nosed species live in grassland or savanna. They differ from the common wombat in having a short-haired rhinarium, which is bare in the common species. Hairy-nosed wombats also have silky, soft fur and large ears compared with the coarse fur and rounded, short ears of the common wombat.

CLASSIFICATION

ORDER MARSUPIALIA
FAMILY VOMBATIDAE

Squirrel Glider

The squirrel glider belongs to one of six Australian possum families collectively known as phalangerids.

CLASSIFICATION

Order Marsupialia
Family Petauridae

Close relatives Seven species of wrist-winged gliders, Leadbeater's possum and four striped possums of New Guinea comprise the family.

Characteristics The distinctive gliding membrane extends from the wrists (rather than the elbows as in the greater glider) to the ankles. The family are grouped according to their dentition and all have a prominent, dark back stripe extending along the forehead. Squirrel gliders nest in family groups, in a round tree hollow, lined with leaves.

Food This glider mainly eats beetles and caterpillars, as well as eucalypt sap, nectar and pollen.

PETAURUS NORFOLCENSIS
Thrust by its hindlegs, the squirrel glider leaps from a tree, spreading the gliding membranes and steering by curving the left or right side. It brings its hindlimbs in towards the body and lands with an upward thrust on a tree trunk.

Young Two young develop in the pouch in 70 days and are then moved to a leaf-lined nest in a tree hollow. Males mark their territory with scent glands.

Habitat The animal is rare due to the clearing of sclerophyll forests and the loss of tree hollows essential for squirrel gliders. It inhabits the coastal regions of much of eastern Australia.

Striped Possum

The striped possum belongs in the same family as the squirrel glider, although it lacks a gliding membrane. These possums are grouped in a genus for their sharp, lower incisors, striped backs and furry, prehensile tails.

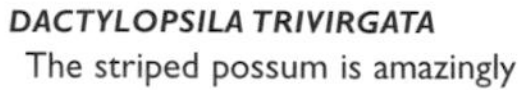

DACTYLOPSILA TRIVIRGATA
The striped possum is amazingly agile, leaping between trees. Although fast, it can be followed because of its noisy rustling, snorting, scratching and slurping.

Close relatives Four striped possum species, seven wrist-winged gliders and Leadbeater's possum comprise the family.

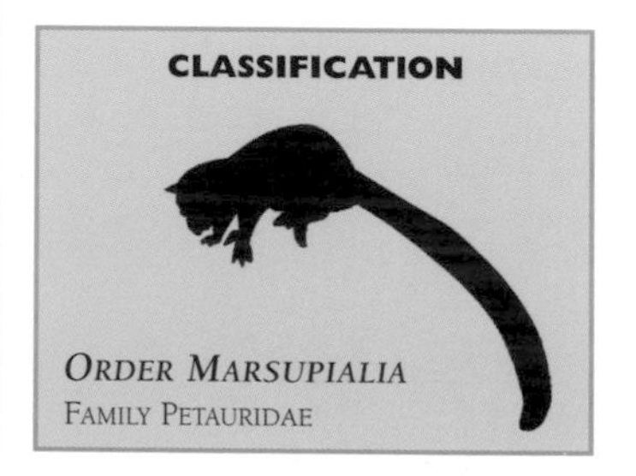

Characteristics The striped possum is a slightly built, skunk-like animal, with a unique, extremely long fourth finger with a powerful claw. It leaves a pungent odor in its wake. It sleeps during the day in a leafy nest inside a tree hollow, and despite its conspicuous coloration is shy and little known.

Food It is an insectivore that feeds on adult and larval insects that burrow in trees, extracting them with the long forefinger. Leaves, fruit and small vertebrates are also eaten.

Young Copulation is brief and raucous. Up to two young are born, attaching to the two teats in the pouch.

Habitat The striped possum is sparsely distributed in far north-eastern Australian rainforests and woodlands, but is more common in New Guinea.

Common Spotted Cuscus

Cuscuses and their other family members are excellent climbers, with the first two digits of the forefeet opposable to the other two and a prehensile tail. Phalangerids have a short face, with the eyes directed forward and a prominent nose.

Close relatives Brushtail possums, the monotypic scaly-tailed possum and 16 cuscuses comprise the family.

Characteristics The cuscus has dense, woolly fur and the males have white blotches. Its ears are almost invisible and with a rim of reddish skin around the eyes, it has been misidentified as a monkey. It is a cautious climber, with the last two-thirds of its tail being prehensile. It is predominantly arboreal, but capable of traveling long distances overland. The mainly nocturnal spotted cuscus sleeps in the rainforest canopy.

Food It prefers tropical fruits, flowers and certain rainforest leaves, but large canines suggest carnivory.

Young The cuscus is mainly solitary, probably with an extended breeding season. There are four teats in the pouch, but usually only one young at any time.

Habitat This species lives on the tip of far northern Australia, but is widespread in New Guinea.

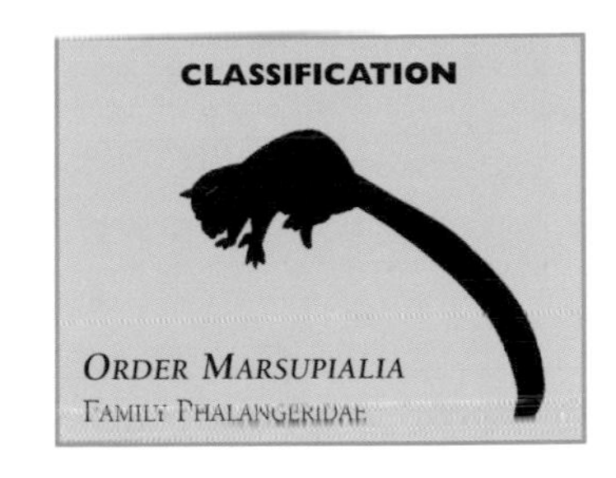

SPILOCUSCUS MACULATUS
Although likened to sloths as tropical canopy dwellers that do not use dens, with a low metabolic rate and thick insulating fur, the cuscus is much more active.

Goodfellow's Tree Kangaroo

Wallabies, kangaroos and tree kangaroos share a family. All but the rock wallabies and tree kangaroos possess powerful hindlimbs with a narrow hindfoot and long fourth toes. Tree kangaroos have markedly shortened feet, and rock wallabies differ in the structure of the pads on their soles. The forelimbs of all members of the family are small and capable of only a simple grip.

DENDROLAGUS GOODFELLOWI
Why kangaroos returned to the trees is an enigma, having evolved from terrestrial kangaroos. In adapting to climbing, their forelegs became longer and stronger, while the hindfeet became shorter and rectangular. It may simply represent a vertical step to make use of a food resource.

Close relatives Kangaroos are collectively known as macropods.

Characteristics Goodfellow's tree kangaroo is distinguishable from others in its genus by its smaller size and color. It is nocturnal and spends most of its time in the canopy, often in pairs. It can hop rapidly, but the group are the only kangaroos able to move their hindlegs independently of each other, which is a useful adaptation for moving along tree branches.

Food The tree kangaroo eats leaves, roots and fruits, especially figs. It can grasp a twig and pull it toward its mouth and hold fruit in its forepaws.

Young Tree kangaroos copulate on the ground, as they are quite awkward in trees. There are four teats in the pouch and one becomes enlarged for attachment of the single young before birth.

Habitat Living in rapidly disappearing oak forest in New Guinea, this species is critically endangered.

CLASSIFICATION

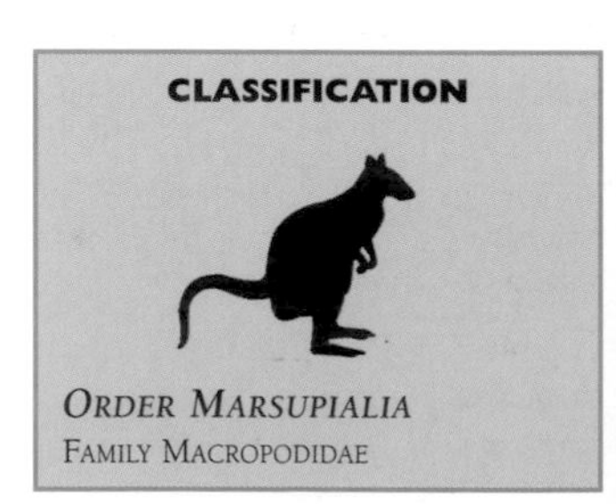

ORDER MARSUPIALIA
FAMILY MACROPODIDAE

Musky Rat-kangaroo

These macropods are separated from the macropodidae in their dentition. They have a shorter hindfoot and a slightly prehensile tail.

Close relatives Rat-kangaroos, potoroos and bettongs comprise the family group.

Characteristics The musky rat-kangaroo is the smallest (head and body length 9 inches or 23 cm) and a very unusual macropod. It bounds rather than hops and with the aid of an opposable digit on the hindfoot, occasionally climbs trees. It forages by day in leaf litter and sleeps at night in a nest of leaves.

Food The species is omnivorous, eating insects, fruit, large seeds and fungi.

Young Unlike other macropods, it usually bears twins that remain in the pouch for 21 days and are then nurtured in large nests of leaves.

CLASSIFICATION

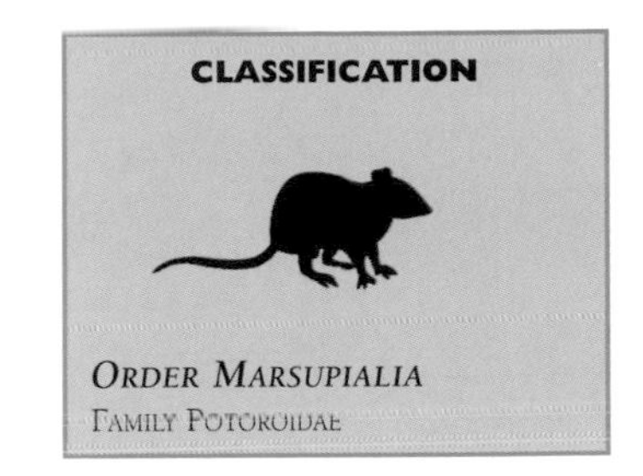

ORDER MARSUPIALIA
FAMILY POTOROIDAE

Habitat The musky rat-kangaroo is common in its limited Australian rainforest habitat, but its range is reducing as a result of logging.

HYPSIPRYMNODON MOSCHATUS
The most possum-like of the kangaroos, this species has a scale-covered, somewhat prehensile tail, which it uses to carry and push nesting material. There is less difference between the length of the fore- and hindlimbs than in "true" kangaroos.

Nailtail Wallabies

The three small Australian nailtail wallabies belong to the same family as kangaroos and other wallabies, but are placed in a different genus, Onychogalea, due to a small horny spur at the end of their tails. Since European occupation, however, the crescent nailtail wallaby has gone extinct, and the bridled nailtail wallaby is rare, though the northern nailtail wallaby is common.

***DORCOPSIS* SP.**
The New Guinean genus of forest wallabies have fine, short fur, with most of the tail densely covered in fine hairs. The tip appears naked and is covered in coarse scales. In these and Australian nailtails, the tip is sometimes pressed vertically into the ground, but no function is known.

Close relatives The wallabies are related to the kangaroos and tree-kangaroos.

SMALL KANGAROOS
Wallabies are generally distinguished from kangaroos by size, wallabies not usually exceeding 44 pounds (20 kg).

Characteristics These are small wallabies, distinctively marked with a mid-dorsal stripe and nail on the tail tip. The claws of the forefoot are used to scrape below vegetation to make a nest for daytime rest.

Food Nailtails are selective feeders on herb foliage, succulents and fruit.

Young Nailtails are usually solitary, but may feed in groups, and have a pouch-young as well as a young at foot.

Habitat Nailtails shelter during the day in open woodlands and grasslands with scattered trees.

CLASSIFICATION

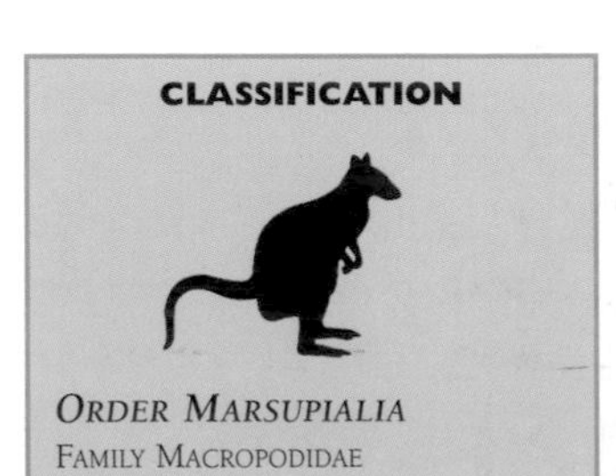

ORDER MARSUPIALIA
FAMILY MACROPODIDAE

Red Kangaroo

Kangaroos and their kin, collectively known as macropods, are characterized by powerful hindlimbs, long hindfeet and a very large fourth toe. The red kangaroo is the largest marsupial, weighing up to 190 pounds (85 kg).

Close relatives The kangaroos, wallabies and tree-kangaroos share a family.

Food This species depends on green herbage for food.

MACROPUS GIGANTEUS
The eastern gray kangaroo is distinguished by its hairy muzzle. Unlike the red kangaroo, females do not become fertile immediately after birth, but after the young departs the pouch.

Characteristics This species' coat is red to blue-gray, and is distinguished from other kangaroos by its white underparts. The muzzle is naked and sharply delineated.

Young The "big red" is intensely sensitive to drought and females may cease ovulating until a drought breaks. Half of the pouch-young die after only two to three months of drought. This protects the reproductive female from the load of suckling a growing young for eight months during drought.

Habitat This is the only truly arid adapted kangaroo and is abundant across most of inland Australia.

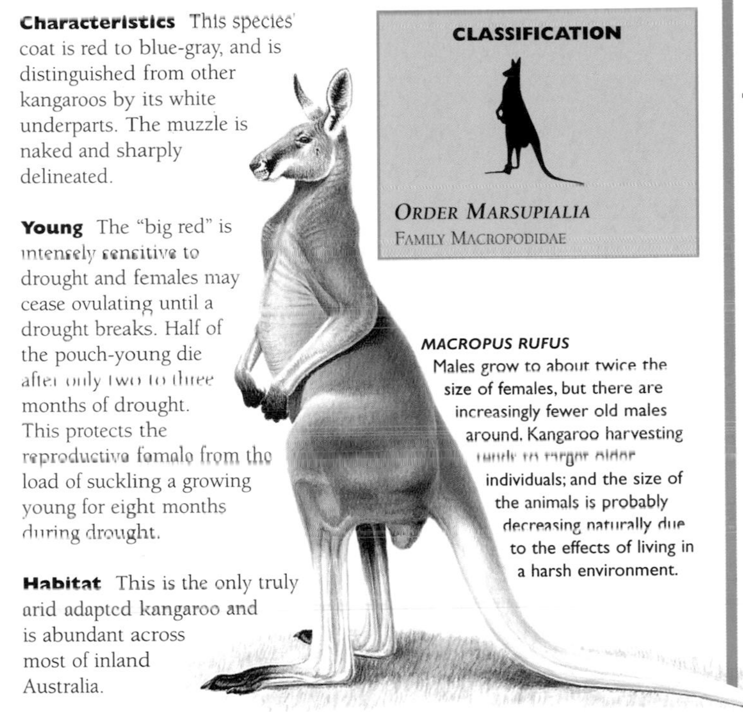

CLASSIFICATION

ORDER MARSUPIALIA
FAMILY MACROPODIDAE

MACROPUS RUFUS
Males grow to about twice the size of females, but there are increasingly fewer old males around. Kangaroo harvesting tends to target older individuals; and the size of the animals is probably decreasing naturally due to the effects of living in a harsh environment.

KANGAROO BEHAVIOR

The hopping gait of kangaroos probably evolved from the bounding style that has been retained in the small and primitive musky rat-kangaroo. The hindlimbs became increasingly larger than the forelimbs and the hindfoot became longer, providing a very effective means for fast, hopping locomotion. But such specialization is not without cost: a kangaroo cannot walk.

Only tree-kangaroos are able to move each hindleg independently when walking along branches, but when swimming, terrestrial kangaroos do kick their legs alternately. The tail is not very flexible, but aids in balancing when hopping.

Grass-eaters Kangaroos manage the digestion of tough grasses with teeth specialized to crop and finely chew. The food is then retained for microbial digestion in a compartment of the stomach.

HOPPING GAIT

Kangaroos move at speed on their hindlimbs by hopping, and slowly with a shuffling motion, taking the weight on the forelimbs and swinging the hindlimbs forward, and always, together.

Front feet and head first.

Twisting so the head is at the pouch bottom.

Finally turning to see out and be ready to jump.

POUCH SAFETY

Kangaroos have a deep, forward-opening pouch, and long after the young is able to leave the pouch, it will return to travel and sleep.

Courtship Males vie for the attention of a female by staying in her vicinity and approaching her during the days before estrus. They will sniff the female's pouch and cloaca, paw at her head and grab her tail. Mating usually occurs straight after a birth, but the young at the teat will have the effect of stopping development of the newly fertilized egg.

Permanently pregnant The stage of development of the fertilized egg is called the blastocyst; and the phenomenon where it is in "suspended animation" is called "embryonic diapause." The blastocyst remains in this state until the older sibling is ready to leave the pouch, at about seven months. If the pouch-young does not survive, the blastocyst starts growing. So a female kangaroo will usually be carrying one blastocyst, one pouch embryo and one suckling young.

Big red The red kangaroo has a reproductive adaptation to drought. The female will stop producing eggs until green grass sprouts. Some males will also stop producing sperm, the number directly related to the duration of the drought.

MALE COMBAT

Competition between males during breeding season may lead to pushing and kicking jousts. Aggressive behavior is accompanied by a gutteral coughing. The court-ship ritual involves showing increasing interest over a period of days.

KINDS OF MAMMALS

ANTEATERS AND ARMADILLOS

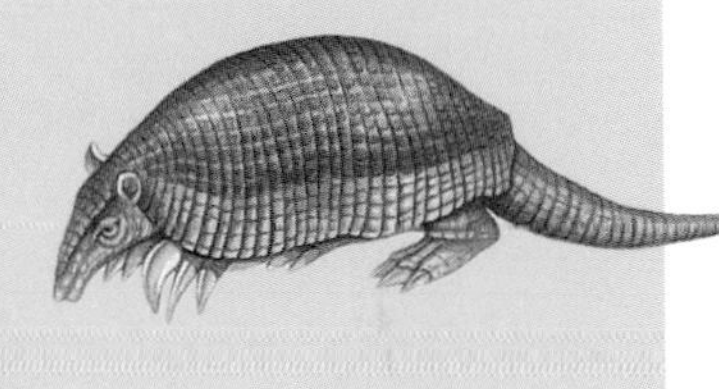

It is difficult to imagine the similarities between heavily armored, fast-moving armadillos; lethargic, tree-living sloths; and the elegant movements of anteaters, which live on the plains as well as in trees. Yet the three living families in the order Xenarthra share a common origin, as well as many physical features. They originated in North America and crossed to South America at the beginning of the Paleocene epoch. The land bridge was severed for at least 70 million years, enabling this bizarre group of animals to evolve in isolation. Four main lines rose from this ancient stock: the modern anteaters, sloths and armadillos, and a burrowing omnivore group that are now extinct.

Giant Anteater

Anteaters have developed astonishing evolutionary adaptations for their specialized diets. The giant anteater is the only member of its family that is not predominantly arboreal; the other three smaller species have prehensile tails.

Close relatives The giant, silky and two species of collared anteaters comprise the family.

Habitat The savannas and open woodlands of South America.

CLASSIFICATION

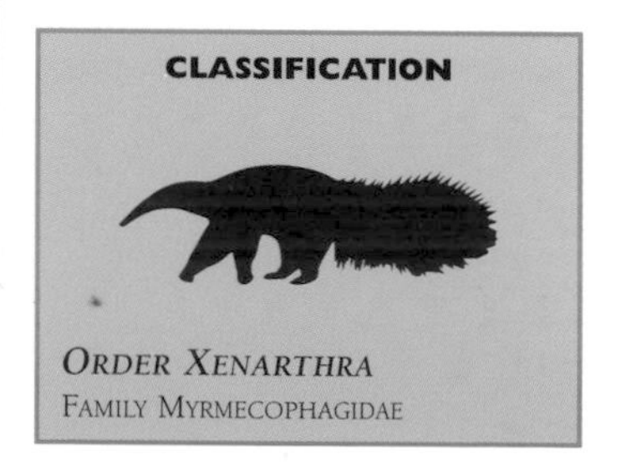

ORDER XENARTHRA
FAMILY MYRMECOPHAGIDAE

Characteristics The giant anteater is the largest xenarthran, with a head–body length of 40–47 inches (105–120 cm), and weighing 44–88 pounds (20–40 kg). It has a highly attuned sense of smell and constantly sniffs the air with its elongated snout. The non-retractile claws compel the anteater to walk on the edges of the feet in a characteristic rolling (though speedy when required) gait.

Food It has a muscular stomach, able to digest the toughened exoskeletons of ants and termites.

Xenarthran distribution

This order is strictly confined to the New World.

MYRMECOPHAGA TRIDACTYLA
Upon locating a potential feeding place, the giant anteater digs an entrance into the rock-hard termite mound with rapid movements of its toughened claws. The long, worm-shaped tongue is protrusible and covered with a sticky saliva secreted by enlarged glands at the base of the neck.

Three-Banded Armadillo

Armadillos are among the smallest and most endangered xenarthrans. The lesser fairy armadillo has a head–body length of 5–6 inches (12–15 cm) and the giant armadillo is comparable in size to the giant anteater. The "armor" develops from skin and is made of strong, bony plates overlaid by horn.

Close relatives There are 20 species of armadillos.

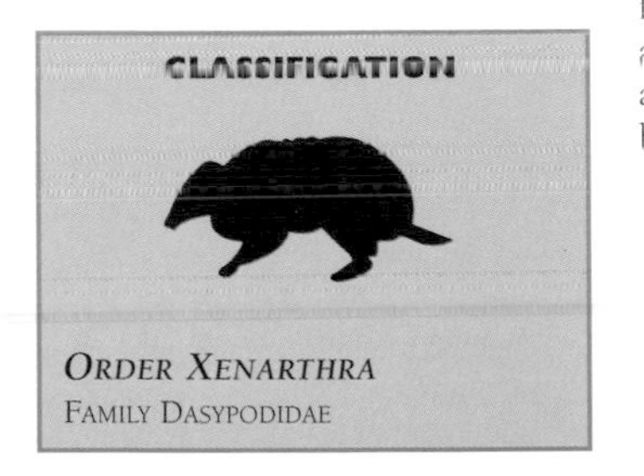

Classification

Order Xenarthra

Family Dasypodidae

Characteristics The armadillos are the most widespread xenarthrans, due to their flexible reproductive behavior and diet. They are generally nocturnal, spending the day curled up with head and tail folded neatly over the belly. All armadillos have one horny shield over the shoulders, another over the hips, but a varying number around the waist.

Food Primarily feeding on insects, they also eat invertebrates, small vertebrates and vegetable matter.

Young For many in the family, if conditions are favorable they are able to breed.

Habitat Most live in semi-arid and even desert zones, though the giant armadillo prefers forest. The nine-banded armadillo is the most widespread species, found from North America to Argentina.

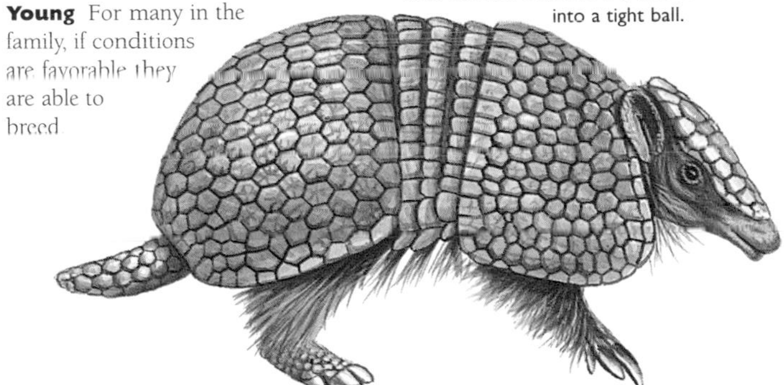

TOLYPEUTES TRICINCTUS
Although at first appearance the toughened carapace would seem to provide adequate deterrent to predators, the armadillo is, in fact, quite vulnerable. The usual defensive reaction is to flee or burrow. Only the three-banded armadillo does not burrow, since it is able to roll into a tight ball.

Kinds of Mammals

Insectivores

With over 400 species, the order Insectivora is a diverse group that shares the specialization of insect-eating. There are six families: solenodons; tenrecs and otter shrews; golden moles; hedgehogs and moonrats; shrews; and moles and desmans. They are all small and highly mobile animals, with long, narrow, often elaborate snouts and teeth specialized for their diet. Fossil evidence indicates that the first primitive placental mammals were insectivores, and their descendants—together with two other insectivorous orders: Scandentia (tree shrews); and Macroscelididae (elephant shrews)—comprise this order.

Greater Moonrat

This family consists of the well-known, spine-bearing European hedgehog and the much less known moonrats or gymnures of South-East Asia. Moonrats are named for their nocturnal habits and rat-like tail rather than any affinity with rodents.

ECHINOSOREX GYMNURA
The greater moonrat weighs up to $4\frac{1}{2}$ pounds (2 kg) and looks quite ferocious with its impressive, open-mouthed gestures.

Close relatives There are six species of moonrats in the family.

Characteristics Moonrats have coarse hair and long tails, unlike their spiny relatives. Their primary mode of communication is olfaction (the sense of smell). This feature is keenly developed in the greater moonrat, which has anal scent glands that exude, to human senses, a foul odor. When placed in strategic parts of the animal's domain, this indicates prior occupation, and may identify its sex and age. Largely nocturnal and terrestrial, moonrats rest at daytime in hollow logs or in empty holes. They hunt in water.

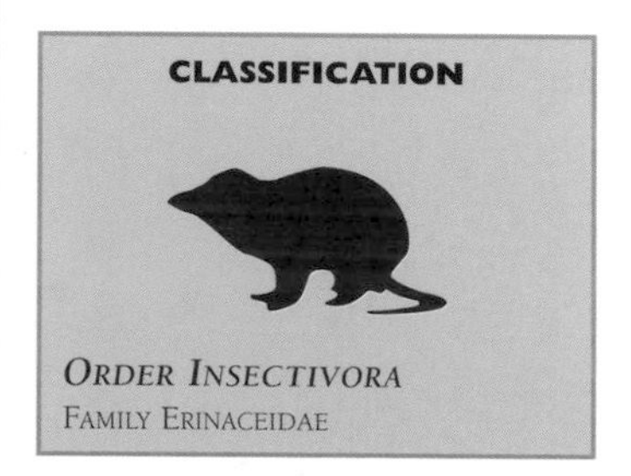

Food The least insectivorous of the order, they feed on insects, frogs, fish, crustaceans and mollusks.

Young The sexual maturity of an animal may be communicated by olfaction.

Habitat Moonrats inhabit lowland areas, including mangrove swamps, rubber plantations, and primary and secondary forests. They require places of refuge for rest during the day and prefer wet areas for food. Moonrats are confined to South-East Asia.

European Hedgehog

The most distinctive characteristic of the hedgehog is its dense coat of spines.They cover the dorsal surface of the body and are normally laid flat. If the animal is threatened, spines stick out at a variety of angles creating a formidable defense system. They curl up into a tight ball to protect the skin of the underparts.

Close relatives There are 13 species of spiny hedgehogs.

Insectivora distribution

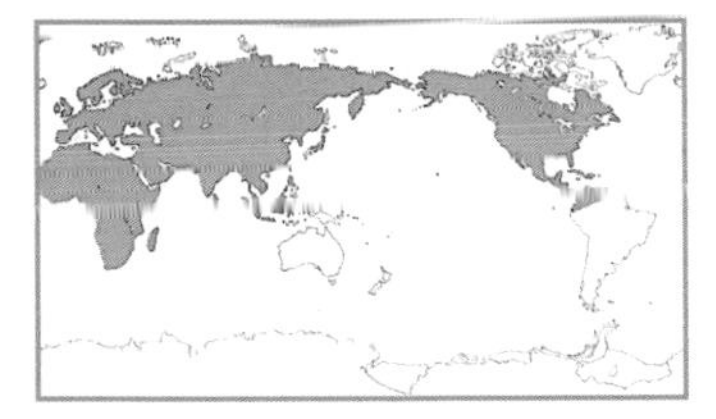

The order is largely confined to northern temperate zones: North America, Canada, Europe, Russia; but also Africa, Southern Asia and northern South America.

Characteristics An adult will have up to 5,000 needle-sharp spines. Hedgehogs are solitary and noctural. They may hibernate to reduce energy expenditure, and construct nests of leaves and grasses.

Food Hedgehogs' favorite food is earthworms, but they also eat other invertebrates, seeds and fruit.

Young In temperate zones, two litters occur per year.

Habitat The European hedgehog is found in deciduous woodlands and even urban gardens, in west and north Europe.

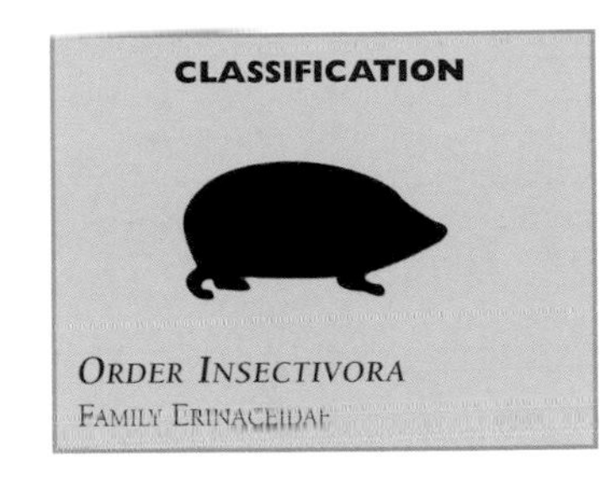

CLASSIFICATION

ORDER INSECTIVORA

FAMILY ERINACEIDAE

ERINACEUS EUROPAEUS

The European hedgehog undergoes a seasonal hibernation in its native land, but forgoes this behavior in New Zealand. It was introduced there in the early 1900s, and has less pressure on it for food resources and low temperatures.

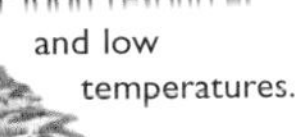

Hedgehog Tenrec

Setifer setosus

The resemblance of the hedgehog tenrec to the "true" hedgehogs is the result of convergent evolution in two distinct families of insectivores. Largely confined to Madagascar, the family has undergone amazing evolutionary radiation, with others resembling moles and otters.

Fossil records indicate that members of this family were well established in Africa 25 mya. Today, the only surviving members of this ancient lineage on the African mainland are the three species of otter shrews, which are confined to small rivers in the tropical forest belt of west and central Africa. The tenrecs were among the first mammals to arrive on Madagascar after its separation from mainland Africa around 150 mya, where they radiated into a wide variety of ecological niches. Various species have fossorial, terrestrial, semi-aquatic and even semi-arboreal tendencies.

Close relatives There are 23 species of tenrecs in Madagascar and otter shrews in west and central Africa, forming the family.

Characteristics Tenrecs are small animals, weighing from 1/2 ounce to 2 pounds (10 g to 1 kg). They are either active at twilight or nocturnal, and most are solitary.

Food Tenrecs prey on invertebrates.

Young Tenrecs retain some primitive mammalian features: males lack a scrotum and, like monotremes and marsupials, females have a cloaca. Interestingly, females produce up to 32 embryos, more than any other mammal, although not all the embryos survive until birth. While usually solitary, some tenrecs appear to form colonies during breeding season.

Habitat Tenrecs inhabit forest and scrub in Madagascar.

Pyrenean Desman

This family is primarily one of burrowing insectivores. Moles are specialized for a subterranean lifestyle; shrew moles forage in tunnels and above ground in leaf litter; but the two species of desman are adapted for a semi-aquatic lifestyle.

Close relatives The family includes 42 species of moles, shrew moles and desmans.

Characteristics Desmans locate their aquatic prey by probing with their proboscis-like snout and clearing debris with sharp, elongated claws. Prey is consumed at the surface where rigorous grooming is carried out to maintain the water-repellent properties of the fur. Desmans construct nests of leaves and dried grasses in the banks of streams.

Food Desmans feed on the larvae of aquatic insects and small crustaceans.

Young A stable pair mate in spring. Males become very protective of their territory, while females concentrate on feeding and gathering nesting material. Gestation is four weeks and the young stay with their mother until seven weeks old, by which time they are already proficient swimmers.

Habitat The Pyrenean desman is confined to fast-flowing streams of the mountain range and parts of northern Iberia.

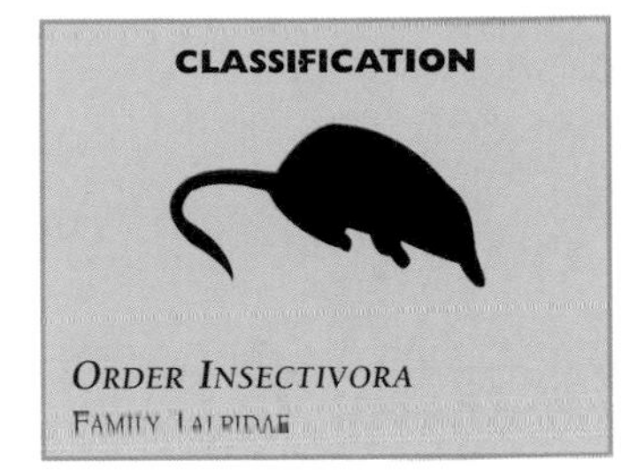

GALEMYS PYRANAICUS
The streamlined body of the Pyrenean desman enables it to glide rapidly through the water, propelled by powerful webbed hindlimbs and steered, to some extent, by a long, broad tail.

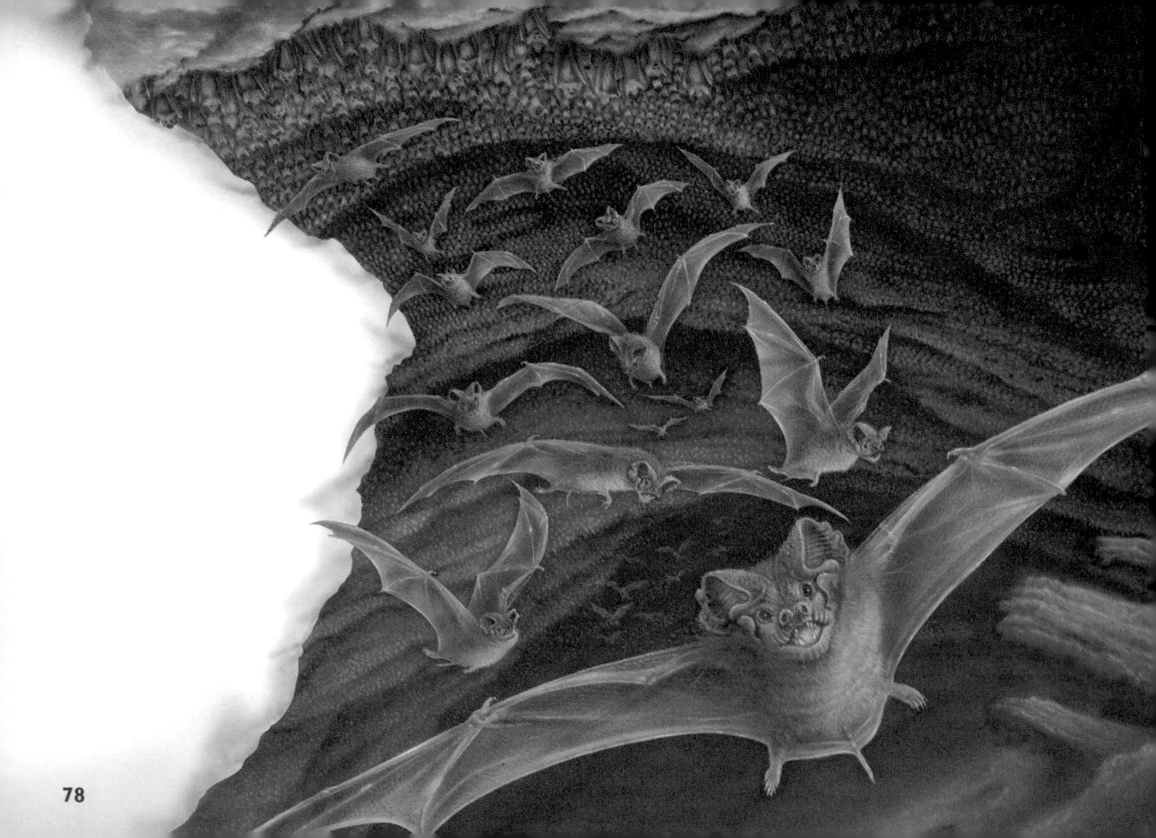

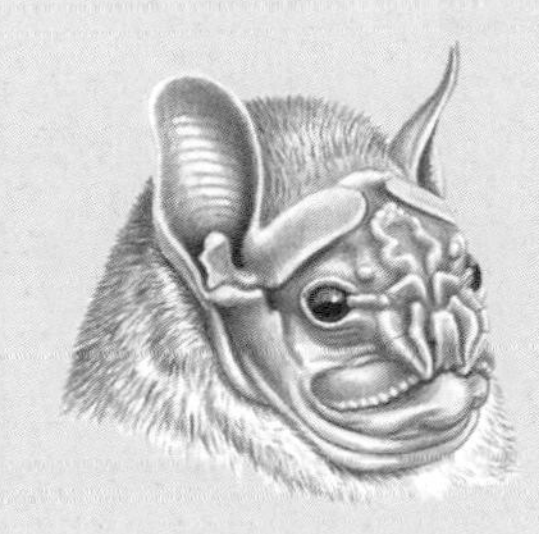

KINDS OF MAMMALS

BATS

Bats are the only mammals capable of powered flight. They originated in tropical forests, perhaps 70 to 100 mya, and appear to have evolved from insectivorous ancestors that developed gliding membranes. The Chiroptera are a large and diverse order, inhabiting all but the coldest environments. Bats have superb hearing and many use high-frequency echolocation for night navigation and hunting. They are classified into two suborders: the Megachiroptera, or Old World fruit bats; and the Microchiroptera, small, mainly insect-eating bats. The megabats have just one family of dog-faced fruit bats and flying foxes, mainly from the tropics. The microbats are varied in their ecology, occurring worldwide.

THE WORLD OF BATS

The wings of bats are their most obvious feature. The bat wing is essentially a modified hand, hence the Greek name Chiroptera meaning "hand wing." The digits, except the thumb, are greatly elongated to support the flight membranes, with the aid of a lengthened forearm. The thumb is usually free of the membrane and has a claw.

VARIED DIET

About 70 percent of bat species feed on insects and other small arthropods like spiders. Most of the rest feed on fruits, flowers, nectar and pollen. The vampire bat (illustrated) is exceptional in that it laps the blood of other animals.

Flight membranes In many bats the flight membranes also join the legs and tail. They are muscular, tough extensions of the body skin. The wings greatly increase the surface area, and have a high concentration of blood vessels, so lose body heat readily. Many behavioral and structural features of bats are geared to dealing with the problems of energy loss.

Energy conservation Flight requires a lot of energy, so bats roost in warm surroundings, or may hibernate or migrate to avoid exposure to the cold. Many bat species can regulate their body temperature, allowing it to fall while roosting during the day. In temperate zones, torpidity is extended into hibernation during winter, when food is scarce, and metabolism is reduced to very low levels. Some temperate species migrate to avoid the extremes of winter. Tropical fruit bats maintain a relatively constant body temperature and roost in humid surroundings, but may follow seasonal fruiting of their foods.

Chiropteran distribution

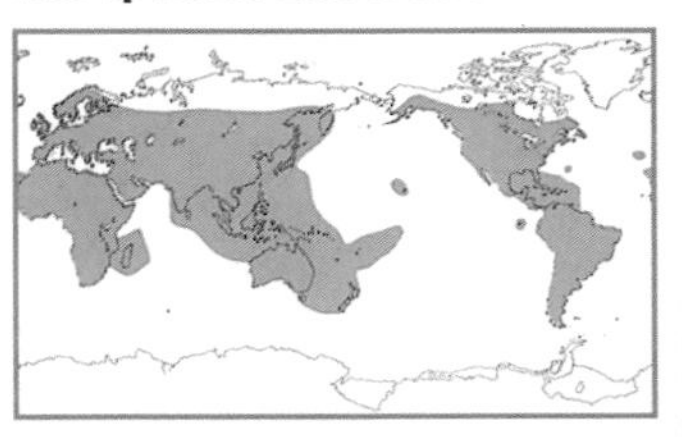

Species are concentrated in the tropics and subtropics, but occur worldwide, apart from: the Antarctic; north of the Arctic Circle; and a few isolated oceanic islands.

EAR ADAPTATIONS

Bats' ears vary widely in size and shape, some with exceptionally long and complex structures. Many also have a tragus, a small projecting flap just inside the ear conch opposite the ear opening, which increases the directional sensitivity of the ear.

Combination of senses Only the Microchiropterans, like the long-eared and tent-building bats (below) make use of high frequency echo-location. Sound reflections are collected in the pinnae, or ear flaps. Although no bat is blind, the insecti-vorous group have very small eyes that are sometimes hidden in the fur.

The eyes of the fruit-eating Mega chiroptera, such as the lesser bare-backed fruit bat (below) are large and adapted to poor light. They have simple, external ears, which lack a tragus. They produce low-frequency orientation sounds by clicking the tongue and other audible calls of many kinds.

The long-eared bat belongs to the widely varied Microchiropteran family Vespertilionidae.

The tent-building bat is a tropical New World leaf-nosed bat of the family Phyllostomidae.

The lesser bare-backed fruit bat is one of a large family of Old World fruit bats, Pteropodidae.

Old World Fruit Bats

The suborder Megachiroptera appear to be an early branch from the bat evolutionary trunk, as they retain some primitive features. The earliest known fossils come from the Eocene (55 mya) of West Germany and North America, but the Old World fruit bats are thought to have originated even as early as 100 mya.

Close relatives The megabats include the Old World fruit bats, flying foxes and dog-faced fruit bats in the single family, Pteropodidae. It includes 41 genera and 63 species.

PTEROPUS POLIOCEPHALUS

The Australian gray-headed flying fox is typical of its family. It has large eyes and excellent vision, and the simple tongue does not extend outside its lips. The long snout accommodates a well-developed organ of smell, employed in selecting food and recognizing other individuals. It eats a varied diet of flowers and fruits and as a result, pollinates many native Australian trees and disperses their seeds.

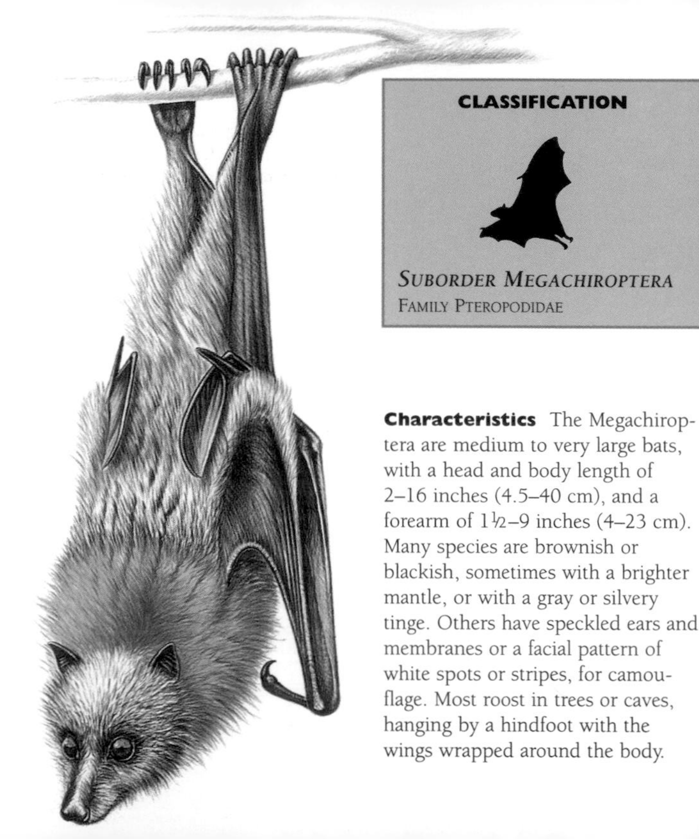

CLASSIFICATION

Suborder Megachiroptera
Family Pteropodidae

Characteristics The Megachiroptera are medium to very large bats, with a head and body length of 2–16 inches (4.5–40 cm), and a forearm of 1½–9 inches (4–23 cm). Many species are brownish or blackish, sometimes with a brighter mantle, or with a gray or silvery tinge. Others have speckled ears and membranes or a facial pattern of white spots or stripes, for camouflage. Most roost in trees or caves, hanging by a hindfoot with the wings wrapped around the body.

Food One of the main classifying features of the megabats is their exclusively frugivorous diet. Fruits, flowers, nectar and pollen of many plant species are utilized, with some plants specialized to facilitate feeding by bats. The digestive system of these bats is not adapted for a fibrous herbivorous diet. Together with fewer and smaller grinding teeth, these features indicate a primitive ancestry.

Habitat The megabats are confined to the tropics and subtropics of the Old World (Europe, Asia, Australia and Africa) where their food source is available throughout the year. Many plants in their habitat have coevolved with the fruit bats, shaped for easy landing, or hanging free within easy reach.

HYPSIGNATHUS MONSTROSUS
The male of the African hammer-headed fruit bat has an enormous muzzle and cheek pouches, which are associated with its loud cries. It produces low, throaty and intense metallic calls to attract females. The female has a head like a flying fox.

EPOMOPS FRANQUETI
The singing epauletted fruit bat has the head shape of a flying fox, and prominent tufts of hair on the shoulders, called epaulettes, which surround a scent gland. It is known to produce metallic, honking calls when navigating at night.

Young Reproduction is generally seasonal, with birth and development coinciding with periods of maximum food abundance. The young of some species are born shortly before the onset of the tropical rainy season. Gestation may be eight months in larger species, usually with one young produced. At birth, baby megachiropterans are relatively large and hairy, and have their eyes open. They cling with strong claws to the mother and suckle for up to three months from one of two teats under the armpits.

New World Leaf-nosed Bats

The suborder Microchiroptera has evolved into such a large and diverse group, that classification is not easily determined by outer physical features. Most feed on insects, but some feed on other mammals, fishes or blood. They are generally smaller than the megabats, but some are larger. Most live in the tropics, but some survive where temperatures drop to below freezing. All microbats make use of high frequency echolocation. The family of New World leaf-nosed bats probably embraces the entire spectrum of bat food habits.

Close relatives The Phyllostomidae is one of 17 families of microbats. It alone comprises 152 species in 51 genera.

Food Few species are exclusively insectivorous, but more or less omnivorous; others are frugivorous.

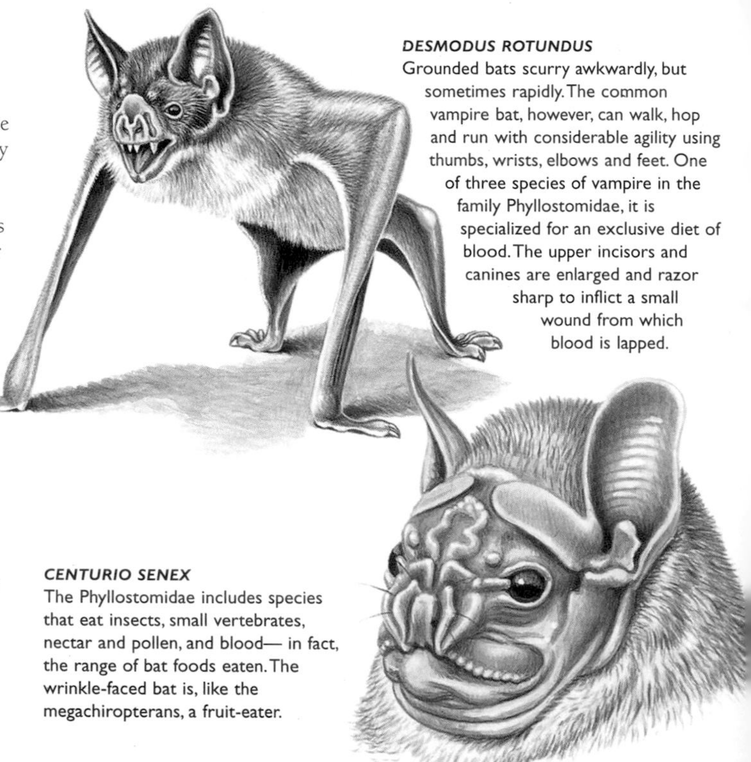

DESMODUS ROTUNDUS
Grounded bats scurry awkwardly, but sometimes rapidly. The common vampire bat, however, can walk, hop and run with considerable agility using thumbs, wrists, elbows and feet. One of three species of vampire in the family Phyllostomidae, it is specialized for an exclusive diet of blood. The upper incisors and canines are enlarged and razor sharp to inflict a small wound from which blood is lapped.

CENTURIO SENEX
The Phyllostomidae includes species that eat insects, small vertebrates, nectar and pollen, and blood— in fact, the range of bat foods eaten. The wrinkle-faced bat is, like the megachiropterans, a fruit-eater.

Characteristics The New World leaf-nosed bats vary in size from small to large: head and body measure 1½–5 inches (4–12 cm); and forearm 1–4 inches (2.5–10 cm). They are named for the generally simple, spear-shaped noseleaf. The ears vary in size and shape, but all have a tragus. Some species have tails and, when present, the tail is usually enclosed in the tail membrane with its tip sometimes projecting slightly beyond the edge of the membrane. Most species are reddish or reddish brown, some with white stripes on the face or back. One species, the white bat, is colored true to its name. The Phyllostomidae exploit a wide variety of roosts from caves to trees, and include three genera that construct a rudimentary shelter. Like all microchiropterans they roost with the wings folded against the sides of the body.

Young Gestation periods in the smaller bats range from about 40 to 60 days, and they are born relatively small and naked, with their eyes closed. The young are left in nursing colonies while the mother forages, although in small species, the young begin to make short flights within three weeks. Nursing lasts from one to three weeks.

Habitat These are mainly forest-living bats of the New World subtropics and tropics.

URODERMA BILOBATUM
The tent-building bat is named for the roosting shelter it makes by cutting through large palm fronds next to the midrib, so that the edges collapse and curl inward. Like other fruit- and nectar-eaters in its family, it has a long muzzle, reduced jaws and teeth, and an extensible tongue, with brush-like papillae.

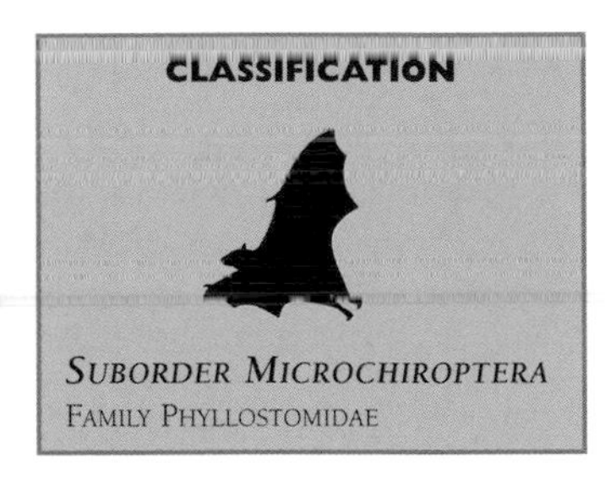

CLASSIFICATION

SUBORDER MICROCHIROPTERA
FAMILY PHYLLOSTOMIDAE

ECHOLOCATION

The use of high-frequency echolocation for navigation, hunting and catching prey is confined to the Microchiroptera. The Megachiroptera lack echolocation of this type, but a few species of the genus *Rousettus* produce orientation sounds that are partially audible by clicking the tongue, and others produce low-frequency calls.

Sound production The sounds are produced in the larynx and emitted through the mouth, or nostrils where there are noseleaves. The noseleaf serves to modify, focus and direct the beam of sound. Echoes received by the ears provide information that can be processed by the brain to provide the bat with an interpretation of its surroundings and the location of its prey. Some moths have a simple ear on the thorax that can detect bat echolocation sounds; others mimic bat calls to interfere with echolocation.

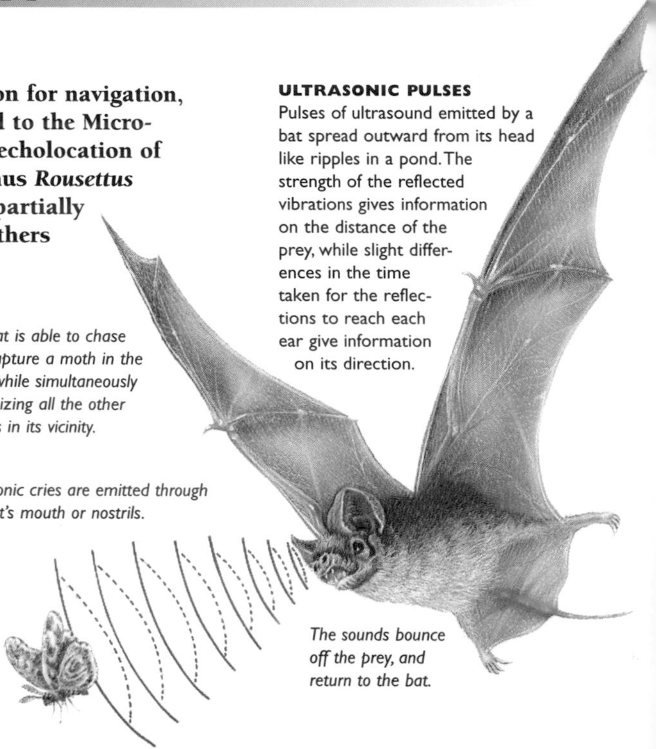

The bat is able to chase and capture a moth in the dark, while simultaneously recognizing all the other objects in its vicinity.

Ultrasonic cries are emitted through the bat's mouth or nostrils.

The sounds bounce off the prey, and return to the bat.

ULTRASONIC PULSES

Pulses of ultrasound emitted by a bat spread outward from its head like ripples in a pond. The strength of the reflected vibrations gives information on the distance of the prey, while slight differences in the time taken for the reflections to reach each ear give information on its direction.

False Vampire Bats

The family is characterized by a conspicuous, long, erect noseleaf; large ears joined at the base, with a prominent tragus; and relatively large eyes.

Close relatives The false vampire bats and yellow-winged bats comprise five species in four genera.

Characteristics False vampire bats are medium to large in size with a head–body length of 2 to 5 inches (4.5–12 cm) and a forearm of about the same length. Although the tail membrane is extensive, the tail is short or absent.

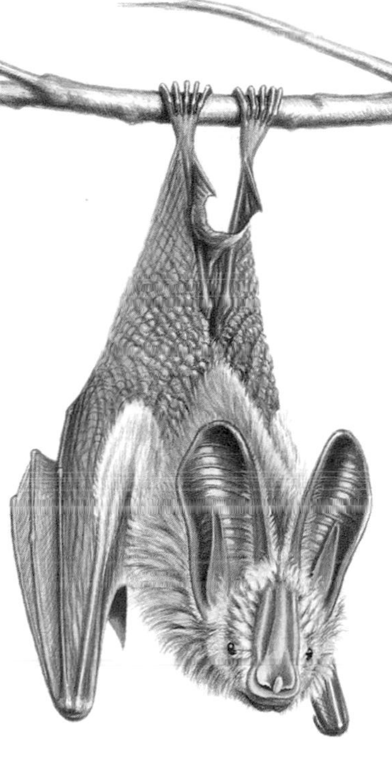

LAVIA FRONS

The African yellow-winged false vampire bat has strikingly colored wing membranes. They are pinkish white in the Australian false vampire or ghost bat (which is the largest microchiropteran); and color varies from blue-gray to gray-brown in other members of the family.

CLASSIFICATION

SUBORDER MICROCHIROPTERA

FAMILY MEGADERMATIDAE

Food The diet includes large insects and small vertebrates, like frogs, birds, rodents and other bats. The bat hangs in wait for passing prey.

Young Gestation is about three months long, and the mothers form nursery colonies with males.

Habitat They roost in caves, rock crevices, hollow trees or foliage in their range in forest and savanna from Central Africa, through South-East Asia to Australia.

Old World Leaf-nosed Bats

Relatives of these bats were inhabitants of limestone caves in the Riversleigh deposits of north-western Queensland, Australia, as early as 25 mya.

RHINONYCTERIS AURANTIUS
Most species of the family are some shade of brown, so the striking golden-colored fur, occasionally with dark brown tips, is a unique feature of the orange leaf-nosed bat. It is illustrated roosting in a typical position for microchiropterans, with wings folded at the side of the body, and head held at nearly a right angle to the back.

CLASSIFICATION

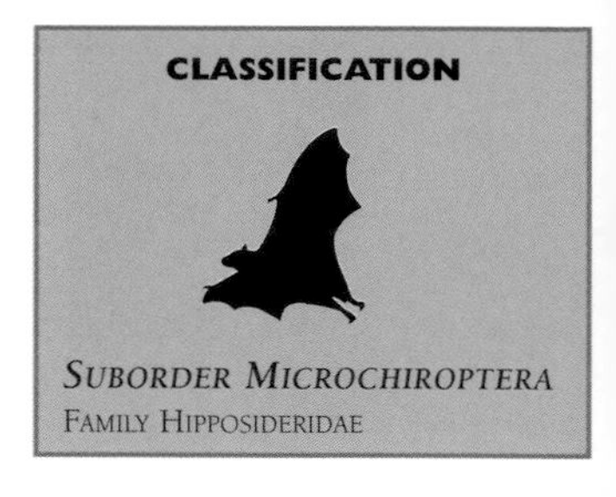

SUBORDER MICROCHIROPTERA
FAMILY HIPPOSIDERIDAE

Close relatives There are nine genera and 66 species in the family.

Characteristics There is wide variation in size among these bats, which have a complex noseleaf, no tragus and moderate-sized ears. The upright part of the noseleaf between the nostrils, the lancet, is rounded in hipposiderids but pointed and triangular in horseshoe bats.

Food These bats eat a wide variety of insects, with moths and beetles often taken in flight.

Young Old World leaf-nosed bats generally roost in caves, in colonies ranging from 20 to several thousand individuals. They may completely abandon a roost if subjected to human interference, but the regular changing of caves may also indicate a seasonal breeding cycle.

Habitat Hipposiderids inhabit the tropics and subtropics of Africa and southern Asia; Vanuatu; and forest and savanna in northern Australia.

Vespertilionid Bats

Vespertilionid bats are so widespread and diverse that the most notable feature of the family is its success: they represent about one-third of bat species.

Close relatives There are 350 species in 43 genera.

Characteristics Species vary in size from very small to large (head– body length of 1–4 inches or 2.5–10 cm). None has a noseleaf and ears are variable with a tragus.

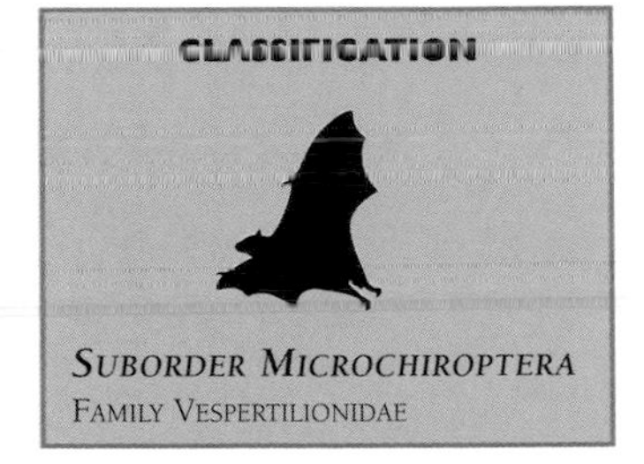

CLASSIFICATION

SUBORDER MICROCHIROPTERA
FAMILY VESPERTILIONIDAE

Food Most species are primarily insectivorous, but a few hunt fish, using long, rake-like feet to capture prey at the surface of lakes or rivers.

Young Many species form enormous nursery colonies and leave the young in the uncomfortably humid roost clinging precariously in pink masses to the cave walls.

Habitat A range of habitats from semi-desert to tropical forest are utilized.Temperate species hibernate or migrate in winter. Mouse-eared bats extend beyond the treeline north and south.

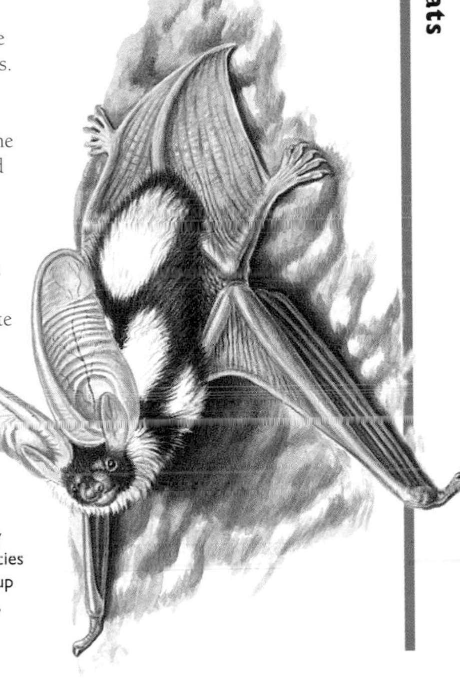

EUDERMA MACULATUM
The spotted bat is notable in the family for its black and white fur, as most species are brown, grayish or blackish.The group usually has an extensive tail membrane, enclosing the tail, which sometimes projects slightly from the edge.

Kinds of Mammals

PRIMATES

The order Primates contains about 200 species of mainly arboreal, keen-sighted, intelligent animals. Several features in combination define the primates. They have hands and feet modified for grasping; nails rather than claws; sensitive, ridged pads on the ends of digits; long gestation periods, life-spans and juvenile dependency; and a large brain. They are classified into two suborders: the Strepsirrhini (lemurs, lorises and bushbabies); and Haplorrhini (tarsiers, New World monkeys, Old World monkeys, apes and humans). The Haplorrhini are further divided on physiological similarities into the Catarrhini (Old World monkeys, apes and humans); and Platyrrhini (New World monkeys).

Indri

This largest of the surviving Madagascan lemurs, rare itself, looks like a typical lemur. Indris are distinguishable from the Lemuridae, however, by their fewer teeth. Their powerful call may carry for over a mile, demarcating territorial limits.

Close relatives There are three genera in the Indridae family: diurnal indris and sifakas, and nocturnal woolly lemurs.

Primate distribution

Most primates inhabit tropical rainforests, but do also extend to temperate to cold environments such as in China and Japan.

INDRI INDRI
Virtually tailless, the indri uses its long hindlegs to propel itself from branch to branch. It spends nearly all its time in trees, fussily searching for tender young leaves and shoots. It sometimes descends to the ground, adopting a hopping gait.

Characteristics The indri is about 28 inches (70 cm) long. It is arboreal and diurnal. Adults seldom engage in any vigorous movement and tend to avoid physical contact with one another, even shunning mutual grooming.

Food Indris eat plant matter.

Young Pairs with young occupy a range of 44 acres (18 ha). Females have one offspring every 2–3 years.

Habitat It is confined to rainforests of north-eastern Madagascar.

CLASSIFICATION

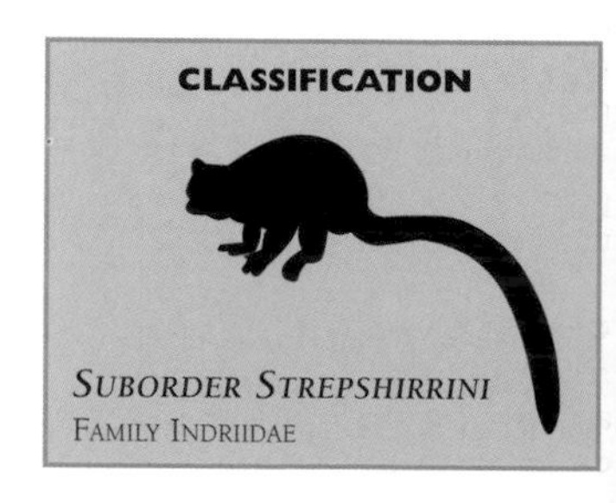

SUBORDER STREPSHIRRINI
FAMILY INDRIIDAE

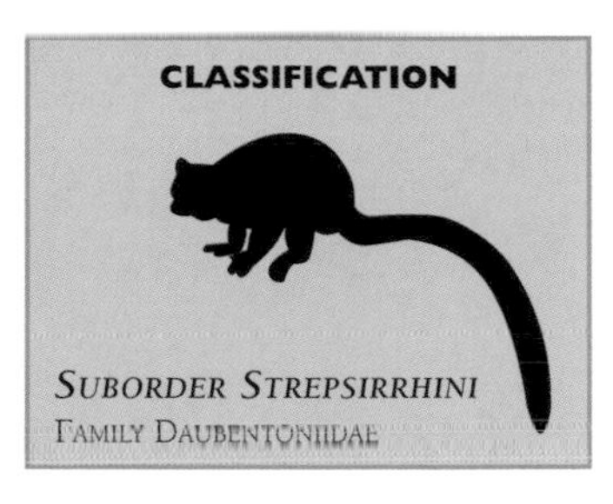

DAUBENTONIA MADAGASCARIENSIS
About half the aye-aye's 31-inch (80-cm) length is bushy tail. These animals have a reputation both for their noisy hunting technique, likened to the sound of woodpeckers, and for their "bad" smell. They use their wire-thin middle fingers to gouge grubs out of crevices, scoop the flesh from fruit and nuts, and to mash food into a palatable paste.

Aye-Aye

The aye-aye is large-eyed, large-eared, coarse-furred and black. It is a shy, solitary, nocturnal animal that sleeps in high treetop nests during daytime. Aye-ayes are on the endangered species list—one reason for their dwindling numbers being that some village communities are prompted by superstition to kill them on sight.

Close relatives It has no living relatives, although a second species appears to have inhabited Madagascar up to less than 1,000 years ago.

Characteristics Big, bat-like ears enable ayes-ayes to hear grubs moving beneath bark. The incisors, reduced to a single pair in each jaw, are huge and open-rooted, so that they continue growing throughout life. These teeth are perfectly adapted for shredding timber and chiseling holes in nuts. Aye-ayes have long curved claws on their hands and feet, except for the opposable, flat-nailed great toes. They lack the dental comb and toilet claw of other Strepsirrhines.

Food Aye-ayes live mainly on wood-boring insect larvae.

Young Aye-ayes sleep and forage alone; males and females do not mix unless the females are on heat. Females bear a single young.

Habitat Aye-ayes are found in dense eastern rainforests on Madagascar and are conserved on a small island.

Gray Mouse Lemur

There are two species of mouse lemurs: the gray and rufus mouse lemur. The gray mouse lemur is one of the smallest primates; it weighs 1–2 ounces (30–60 g) and is 5 inches (12 cm) long. Like most members of the suborder Strepsirrhini, mouse lemurs have a long "toilet claw" on the second digit of the foot and a "dental comb" of six forward-pointing teeth in the front of the lower jaw, which they use for grooming.

MICROCEBUS MURINUS
Mouse lemurs move quadrupedally, scampering through the trees on all fours. They have the large eyes and the long, moist, dog-like snout, called a rhinarium, so typical of most lemur species.

Close relatives Dwarf and mouse lemurs share the subfamily Cheirogalinae.

Characteristics The gray mouse lemur is mainly solitary, though females and young may share their daytime sleeping places. They store fat in their rumps and tails as a food reserve for dry periods when they estivate (that is, pass time in a torpid condition).

CLASSIFICATION

Suborder Strepsirrhini
Family Lemuridae

Food A varied diet includes insects, spiders, birds, eggs, chameleons, tree frogs, sap, fruit, flowers and leaves.

Young Female mouse lemurs ovulate twice a year. Twin offspring are the norm, but occasionally one or three young are born. During anestrus, skin grows over the vagina, effectively closing it.

Habitat Both species are abundant in Madagascar. Gray mouse lemurs thrive in the dry "spiny forest" area of the Strict Nature Reserve.

RED RUFFED LEMUR

Ruffed lemurs belong to the subfamily Lemurinae. The two species of ruffed lemurs—the red and black-and-white—are separated into a genus of their own: Varecia. Ruffed lemurs are among the 29 species of primates classified as endangered.

Close relatives Relatives are the so-called true lemurs of the subfamily Lemurinae.

Characteristics The red ruffed lemur has the typical fox-like face of true lemurs with large eyes and a pointed snout (rhinarium). It has a well developed "toilet claw" and a "dental comb" for grooming. The tails of Lemurinae are larger than in the dwarf and mouse lemurs, and they have prominent cheek whiskers and ear tufts.

Food Ruffed lemurs eat fruit and gorge themselves on ripe figs, consuming large quantities at a sitting.

Young Two or three young at a time are born in nests where they stay for about two weeks. They are then carried in their mother's mouth from one tree to another until they are able to follow the adults.

Habitat The species is confined to the forests of the Masoala peninsula in Madagascar.

VARECIA VARIEGATA RUBRA

The handsomely marked red ruffed lemur is the largest of the five families of lemur. It moves about in quadrupedal fashion. It is nocturnal and usually seen in pairs, but little is known about the behavior of this species in the wild.

CLASSIFICATION

SUBORDER STREPSIRRHINI

FAMILY LEMURIDAE

Ring-tailed Lemur

Fossil deposits indicate the presence of lemurs in North America and Europe in the early Eocene. They disappeared from the Northern Hemisphere during the Miocene, retreating to tropical forests where they did not have to compete so fiercely with more advanced primates. Well named for its striped tail, the ring-tailed lemur is the only species so marked.

Close relatives The subfamily of "true lemurs," Lemurinae, contains four genera.

Characteristics This species is about 3 feet (1 m) long, more than half of which is tail. It is territorial, usually forming troops of 12 to 20 in which females are dominant. Lemurs have scent glands on their forearms and inside upper arms for marking their territory.

Lemur catta
Ring-tailed lemurs live mainly on the floors of open forests. Males have glands on the wrist armed with horny spurs, which they rub onto small trees with an audible "click," leaving a slash in the bark impregnated with scent.

Food It eats fruit and seeds.

Young Females are in estrus for just one day during a two-week mating season in fall (April). After a gestation of 136 days, the single young are born in late winter (August and September). At first they cling to the mother below her belly, climbing onto her back as they grow. The female vagina is closed during anestrus by a flap of skin.

Habitat Ring-tailed lemurs are unique to the dry rocky areas of south-western Madagascar.

CLASSIFICATION

Suborder Strepsirrhini
Family Lemuridae

PHILIPPINES TARSIER

Tarsiers are an intermediate form between lemurs and monkeys. The Philippines tarsier does not look much like a monkey at all, with its long hindlegs, long tarsal region, and long, thin fingers and toes. In the past they have been classified alongside the Strepsirrhini, rather than with monkeys and apes.

Close relatives There are four extant species of tarsiers.

Characteristics In Indonesia, the tarsier is known as "binatang hantu" meaning ghost animal. This name aptly describes their nocturnal way of life and inconspicuous presence in the forest trees, where they are not uncommon. They are unable to move their enormous eyes within the sockets and rotate their heads through 180º to compensate. They form a subgroup of Haplorrhini, separated from the large-brained monkeys, apes and humans.

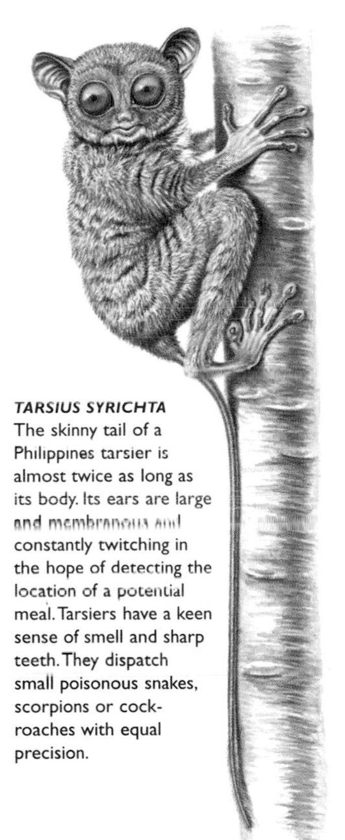

TARSIUS SYRICHTA
The skinny tail of a Philippines tarsier is almost twice as long as its body. Its ears are large and membranous and constantly twitching in the hope of detecting the location of a potential meal. Tarsiers have a keen sense of smell and sharp teeth. They dispatch small poisonous snakes, scorpions or cockroaches with equal precision.

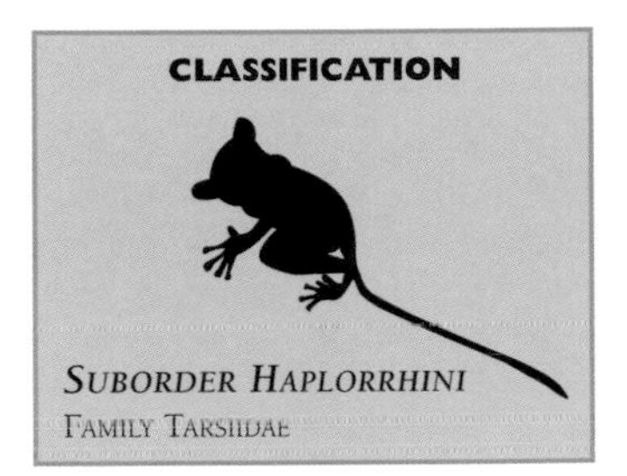

CLASSIFICATION

SUBORDER HAPLORRHINI
FAMILY TARSIIDAE

Food Tarsiers are exclusively carnivorous, never eating plants. They eat insects, lizards and other small vertebrates.

Young Tarsiers live in pairs. Females produce a single, well-furred young with open eyes, after six months' gestation.

Habitat Philippines tarsiers occupy secondary forest on the archipelago of the same name. Other species are found on Borneo, Sumatra and Sulawesi.

OLD AND NEW WORLD MONKEYS

About 40 million years ago, higher primates—monkeys and apes—began to take over from the primitive lower primates —lemurs, bushbabies, lorises and tarsiers. Two groups of monkeys evolved. The New World monkeys are forest dwellers of South and Central America. Old World monkeys from Africa and Asia are ground and tree dwellers.

Old World monkeys Two major groups of Old World monkeys are recognized. The mainly African cercopithecids have simple stomachs and deep food-storage pouches in the lining of the cheeks. These include the guenons, which live in family troops; the herd-living baboons and their relatives, like geladas and mandrills; and macacques. Colobid monkeys are mainly Asian leaf-eaters, with stomachs that have chambers that break down cellulose. These are the langurs, the golden snub-nosed and proboscis monkeys, and relatives.

Old World monkeys have prominent noses with narrow nostrils that face forward.

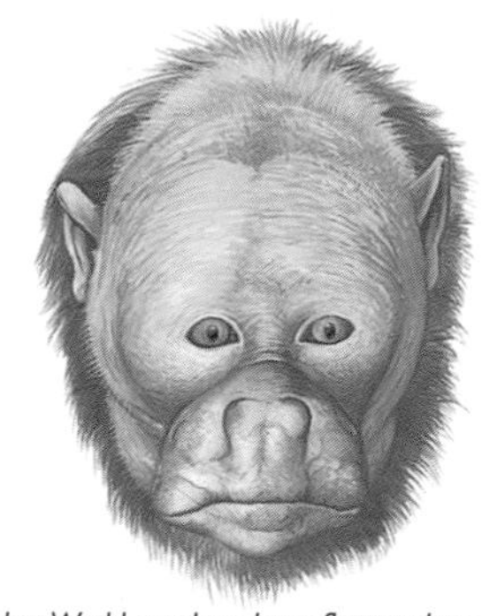

New World monkeys have flattened noses with nostrils that face sideways.

FACIAL COMPARISON
New World monkeys are immediately distinguishable by the nose: the nasal septum is broad, so that the nostrils point sideways. Old World monkeys have thin, forward-facing nostrils that are close together and directed downward. In both groups the nose is dry and sparsely haired, as opposed to the rhinarium of lower primates.

New World monkeys This group is puzzling to paleontologists, because their fossils are 35 million years old, from a time when South America was already separated from Africa, where their closest present-day relatives are found. The pygmy marmoset is the smallest living monkey, weighing 4½ ounces (125 g). Together with tamarins, marmosets are unique among primates having claws on all digits except the great toe, which has a nail. They are classified in the family Callitrichidae. The second family of New World monkeys are the Cebidae. Night monkeys and titis, tend to live in pairs, titis being very closely bonded. Squirrel monkeys live in larger troops. The prehensile-tailed howler monkeys live in troops of 10–30 individuals. Spider monkeys split up into smaller foraging groups of unstable composition from day to day. Others include the sakis and uakaris (unique in having a short tail), intelligent capuchins and endangered woolly monkeys.

Spider monkeys: New World, cebid, monkeys

LIFESTYLES OF THE OLD AND NEW

Spider monkeys are typical New World monkeys, being tree-dwelling herbivores with prehensile tails that live in large troops. The thumbs are not markedly opposable to the other fingers. The proboscis monkey is the largest of the colobids. These Old World monkeys swim well and gallop along tree branches, but the tail is not prehensile. They have hook-like, thumbless hands and walk on all fours. The unmistakable nose of males continues to grow beyond maturity

Proboscis monkey: Old World, colobid, monkey

MONKEY CLASSIFICATION

Old World monkeys
Superfamily Catarrhini
Family Cercopithecidae
Subfamily Cercopithecinae: guenons, baboons, mandrills, macaques, geladas.
Subfamily Colobinae: langurs, colobus, snub-nosed and proboscis monkeys.
New World monkeys
Superfamily Platyrrhini
Family Cebidae: night monkeys, titis, capuchins, sakis, uakaris, howlers, spider and woolly monkeys.
Family Callitrichidae: marmosets and tamarins.

Cottontop Tamarin

Tamarins and marmosets are unique among monkeys in bearing claws on all digits except the great toe; and are distinguished by their dentition and jaw shape. The strikingly crested cottontop tamarin has been indiscriminately collected.

Close relatives Tamarins and marmosets are a subfamily of New World monkeys.

Characteristics This sociable animal lives in family groups formed by pairs and several sets of offspring. Older offspring hang around with the parents and help rear the infants.

Food The cottontop tamarin lives on fruit and secondarily eat insects.

Young Twins and occasionally triplets are born twice a year.

Habitat The rainforest understorey of Colombia and Panama.

SAGUINUS OEDIPUS
In the wild, the cottontop tamarin usually lives in trees in large groups, often with more males than females. Males and older offspring share in care of the young. In captivity, they seem more easily kept in pairs, and there are thriving colonies at several major zoos.

CLASSIFICATION

Suborder Haplorrhini
Family Callitrichidae

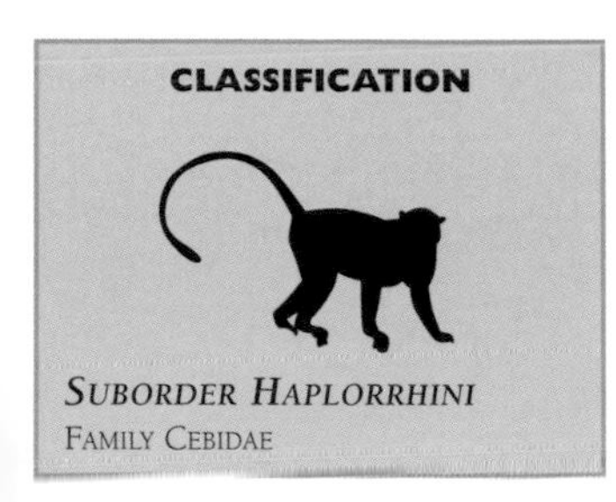

NIGHT MONKEY

Although it was once thought that night monkeys were a single species, five to ten are now recognized. There are obvious variations in color from one area to another; and striking differences in chromosome numbers—between 46 and 56—distinguish the species.

Close relatives Night monkeys, genus *Aotus*, and titis, genus *Callicebus*, are the most marmoset-like New World monkeys.

AOTUS TRIVIRGATUS
Night monkeys are the only entirely nocturnal simians. They live in small family groups that sleep together during the day in hollow tree trunks. Their lack of color vision is no handicap in the dark and they see well enough to be able to leap from tree to tree. Their bushy tails are as long as the rest of them, but are not prehensile.

Characteristics Night monkeys have forward-facing eyes, typical of monkeys and apes, which give them stereoscopic vision. They are pair-living and it is not known if they have extended helper systems. Their heads and bodies are about 10–20 inches (25–50 cm) long, and they have dry noses, sparsely covered in hairs. Their loud voices account for their alternative common name—owl monkeys.

Food Night monkeys feed mainly on fruit, but they supplement their diet with young shoots, leaves, flowers, insects and small animals.

Young Female night monkeys produce one offspring. Pairs seem less closely bonded than titis.

Habitat Night monkeys are widely distributed from Panama southward to the Gran Chaco of Argentina and Paraguay. They are absent from the Atlantic coast forest of Brazil.

Red Howler Monkey

This is a determinedly territorial New World monkey—and the noisiest. Its howl is amplified by a resonating chamber beneath the chin and may carry as far as 3 miles (5 km). Males have the largest chamber and the loudest calls. The red howler monkey lives high in the trees in troops of 10–30, vocally defending its right to a patch of forest.

Close relatives All howler monkeys and other species in the subfamily Cebidae are related.

Characteristics The red howler monkey is the largest of the prehensile-tailed howlers. It is coppery red in color. It moves on all fours in the middle to upper storeys of the forests, seldom in haste. It uses its tail to anchor itself when feeding or at rest.

Food These monkeys are entirely vegetarian and have an enlarged caecum to assist in digesting the vast quantity of plant material they consume.

Young Females produce a single offspring and may give birth at any time of the year.

Habitat Red howler monkeys range across Venezuela.

ALOUATTA SENICULUS
The various species of howler monkeys are differentiated by their color, ranging from all-black, through black-and-gold, black-and-red or brown, to the red of this species. Sometimes there is sexual dichromatism, that is, variation in color between sexes of the same species.

CLASSIFICATION

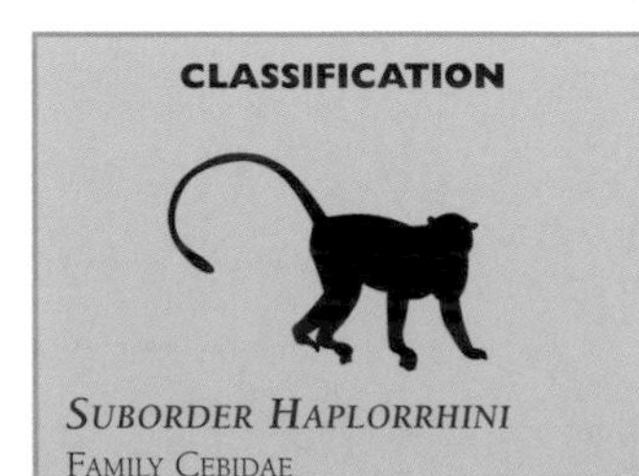

Suborder Haplorrhini
Family Cebidae

Black Howler Monkey

Of the six species of howler monkeys, four may be described as black or predominantly black. The black howler is unusual in being sexually dichromatic—adult males are black; females and young are creamy gray.

Close relatives All howler monkeys and other species in the subfamily Cebidae.

Characteristics Howlers of all species are stoutly built and bearded. The underside of the tip of the tail is hairless to provide a better grip. Howlers rarely descend to ground level. The hyoid bone is enlarged into a resonator that connects with the windpipe. This structure causes the jaws to protrude.

Food Leaves are the howler monkeys' primary food. The diet may be supplemented with fruit.

ALOUATTA CARAYA

The territorial boundaries of family space are a matter of noisy importance. Howlers frequently engage in early morning roaring matches with neighbouring clans. The calls of the mantled howler have been likened to the cheers of a crowd at a football stadium; the red howler's call has been described as the sound of pounding surf or a roaring wind.

Young A single offspring is born at any time of the year. If the milk supply is not forthcoming due to death, sickness or injury of the mother, females may suckle young that are not their own.

Habitat The black howler monkey ranges across a huge area of rainforest from northern Argentina to southern Brazil.

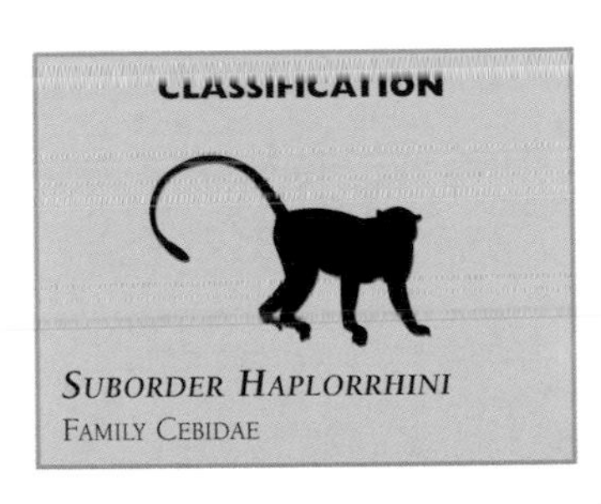

Diana Guenon

Diana guenons are one of about 20 species in the genus Cercopithecus, commonly known as guenons. These Old World monkeys form troops, led by an old male, of as many as 30 animals that forage, feed and sleep together.

Close relatives The subfamily Cercopithecinae includes the guenons and baboons of Africa.

Characteristics Cercopithecus monkeys are a varied and brightly colored lot. In rainforest regions, members of three species-groups live together in a remarkably symbiotic manner. Like other guenons, diana monkeys have long prehensile tails and are excellent climbers. Males are bigger and heavier than females, but otherwise look the same.

Food Diana monkeys mainly feed on leaves, fruit, buds and other plant material but may also eat insects, birds' eggs and nestlings.

Young Females have an ovulatory cycle and bear a single offspring at any time of year. Gestation is about seven months and the infant is suckled for six months.

Habitat Diana guenons are found in the rainforests of West Africa from Sierra Leone to Ghana.

CERCOPITHECUS DIANA
At present, diana guenons are relatively plentiful in the high forests of West Africa, but these monkeys are hunted for their meat as well as threatened by logging. Their habit of moving about in noisy groups in early morning and late afternoon makes them easy targets.

CLASSIFICATION

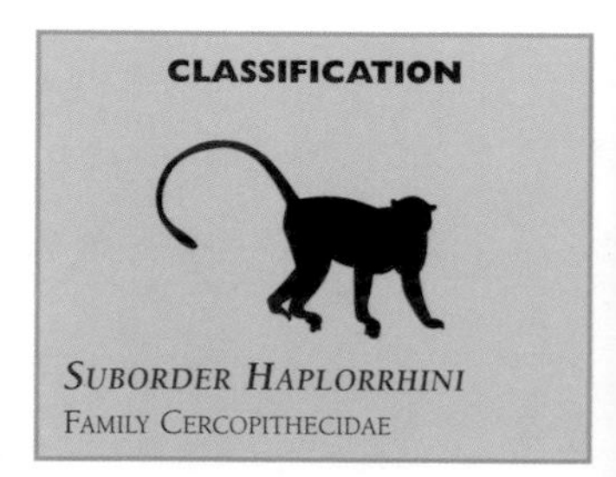

Suborder Haplorrhini
Family Cercopithecidae

Vervet Monkey

These are ground-foraging monkeys, but are adept at climbing, jumping and swimming, using trees as dormitories and safe havens. A dominant male leads a troop of several younger males and females with their young, and may join forces with other troops during the day.

Close relatives This species is most closely related to guenons.

Characteristics The vervet's head–body measurement is 16–31 inches (40–80 cm) long; the tail is 20–27 inches (50–68 cm). Normally relatively quiet, male vervets call harshly when disturbed.

Food The vervet monkey mainly eats vegetable matter, but will also take some insects, spiders, lizards, eggs and nestlings when it can.

CERCOPITHECUS AETHIOPS

Vervet monkeys post sentries to look out for predators, such as leopards, raptorial birds and snakes. Special calls for each type of enemy warn the rest of the troop that danger threatens.

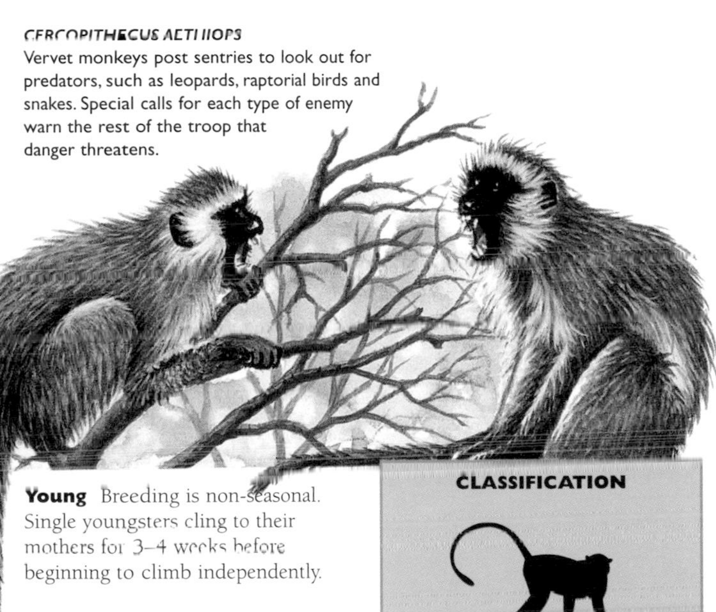

Young Breeding is non-seasonal. Single youngsters cling to their mothers for 3–4 weeks before beginning to climb independently.

Habitat Found on the African savanna and at the edges of woodland from Senegal to Somalia and south to South Africa.

CLASSIFICATION

SUBORDER HAPLORRHINI
FAMILY CERCOPITHECIDAE

MANDRILL

Mandrills are territorial and live in troops headed by a dominant male. They inhabit rainforest but are primarily terrestrial during the day, using trees for sleeping. They are difficult to observe in the wild but seem to go about in troops of between 20 and 50 with one or more mature males in charge.

Close relatives Mandrills are closely related to baboons, geladas, drills, mangabeys, talapoins, swamp monkeys and macaques.

Characteristics Mandrills have bare faces and buttocks and a stub of a tail. The cheeks are ridged; the eyes close-set and sunken. Stout of body, males may weigh 55 pounds (25 kg); females weigh less than half.

MANDRILLUS SPHINX

The mandrill is also known as the forest baboon. Male mandrills are brightly colored. They have a remarkable face with bony swellings along the sides of the red nose. The markings of females and juveniles are more muted.

Food Mandrills are omnivorous and eat fruit, seeds, leaves, nuts, fungi, roots, insects, small reptiles and amphibians.

Young Female mandrills give birth at any time of year but more babies are born from December to February than in other months.

Habitat Mandrills are found in the rainforests of equatorial Africa, including Gabon, south Cameroon and the Congo.

CLASSIFICATION

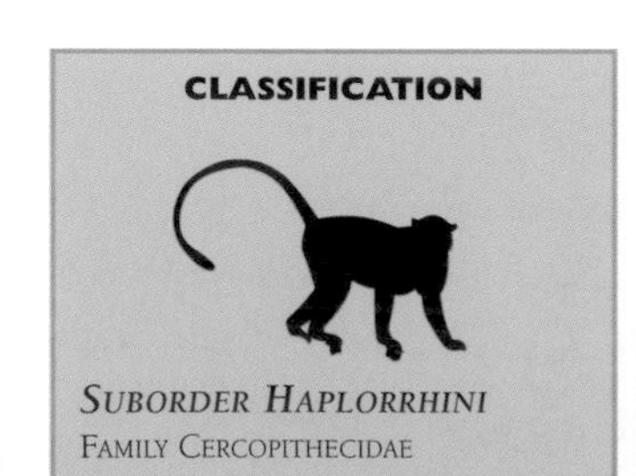

SUBORDER HAPLORRHINI
FAMILY CERCOPITHECIDAE

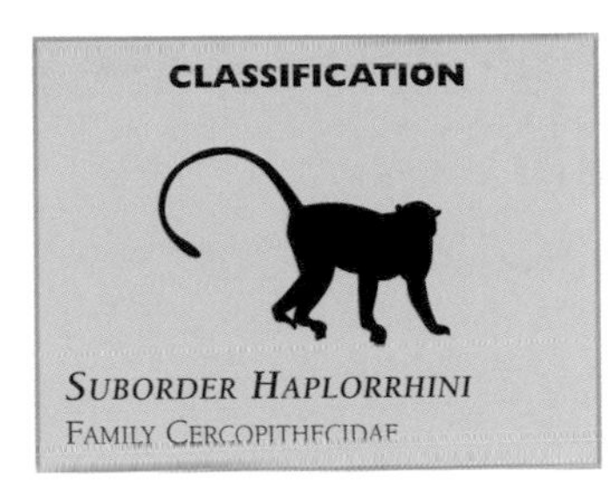

CLASSIFICATION

SUBORDER HAPLORRHINI

FAMILY CERCOPITHECIDAE

THEROPITHECUS GELADA
Like other cercopithecines, geladas have cheek pouches that open beside the lower teeth and extend down the side of the neck. They are able to cram these quickly with food—about a stomach's load—and then scamper away from competitors or danger to eat safely.

GELADA

Geladas are the only survivors of an important fossil group. Although sometimes called gelada baboons, they vary markedly from genus Papio in visual and vocal communication patterns.

Close relatives Geladas are a single species, but baboons and all the African cercopithecids are their closest relatives.

Characteristics The large, baboon-like gelada shuffles round on its haunches plucking plant material with its opposable thumb and index finger. Fatty pads on the buttocks ensure that this movement is not too uncomfortable. Adult males have a long mane and both sexes have a large patch of hairless red skin on the chest.

Food Grass, seeds, roots, bulbs and fruit are varied with insects.

Young Most young are born singly between February and April, though twins occasionally occur. Females suckle their young for up to two years and do not come on heat again for 12–18 months after giving birth.

Habitat Geladas are restricted to mountains, rocky ravines and alpine grassland above 6,500 feet (1,980 m) in Ethiopia.

Sulawesi Crested Macaque

The macaque species of Sulawesi, of which there are about seven within the "silenus-sylvanus" group of macaques, have developed in isolation since before the last Ice Age when their island home was separated from Asia and Australia. Sulawesi crested macaques live in fairly large hierarchical groups presided over by a dominant male.

Close relatives Sulawesi crested macaques are related to the other species on the island, to other macaques, and to baboons, drills, mandrills, mangabeys, talapoins and swamp monkeys.

Characteristics As its Latin name suggests, the fur of this heavily built, tailless animal is entirely black; its rump is pink. Male Sulawesi crested macaques are larger than females, averaging a head and body length of 21 inches (55 cm) and weighing 22 pounds (10 kg).

MACACA NIGRA

What the Sulawesi crested macaque lacks in color and tail, it makes up for in crowning glory—the hair on the head rises to a stiff crest. Despite being protected by Indonesian law, the macaques of Sulawesi are hunted for food and trapped for the pet trade.

CLASSIFICATION

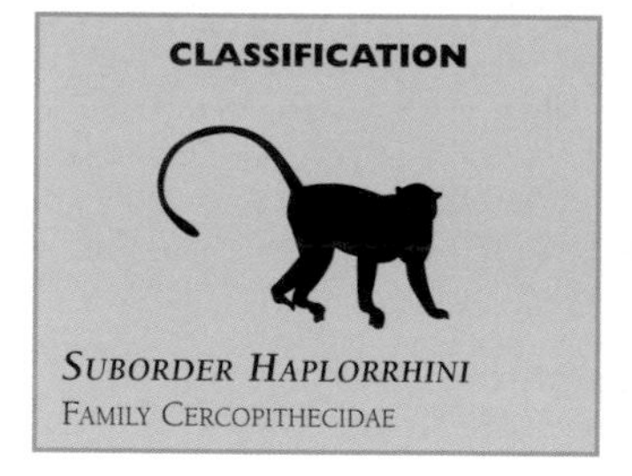

Suborder Haplorrhini

Family Cercopithecidae

Food These monkeys are omnivorous.

Young Females have a cyclic pink perineal swelling. Infants have a soft natal coat that is replaced after about two months. Adoptive suckling of infants has been observed.

Habitat Sulawesi crested macaques, also known as Celebes macaques, reach their greatest numbers in the dense forests on Sulawesi, the third largest island in the Indonesian archipelago.

HAMADRYAS BABOON

PAPIO HAMADRYAS

Hamadryas baboons live in hierarchically structured social groups. For as long as he is fit, one older male leads a harem of females and variously aged offspring on daytime foraging. At night, several hundred hamadryas baboons hole up against predators in rocky outcrops or in trees. These animals can cope with a harsh environment providing they have access to water. When it is not readily available, they may dig to find it.

A silvery gray coat about the shoulders and red naked skin on face and perineum are marks of maturity in hamadryas baboons. Females and juveniles are olive-brown.

Close relatives These include other baboons and geladas, and related African cercopithecids.

Characteristics Hamadryas baboons are terrestrial animals. The males weigh around 37 pounds (17 kg), about twice the size of the females. Both sexes have a long, dog-like muzzle. They mate with savanna baboons where their ranges overlap in northern Ethiopia.

Food These baboons mainly subsist on grass seeds, roots, bulbs and other plant matter, but they are opportunistic feeders and relish insects and small animals when they come across them.

CLASSIFICATION

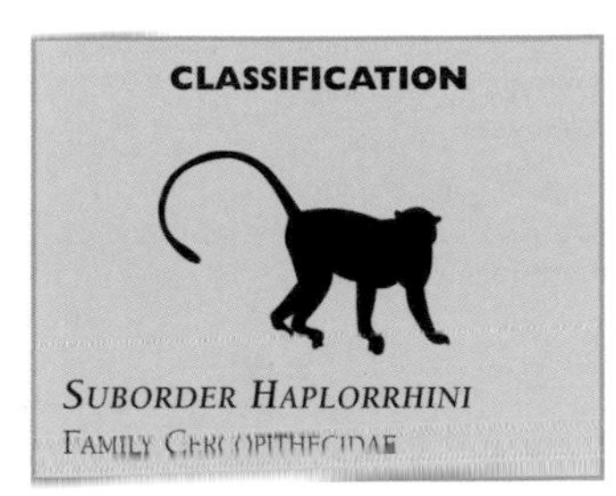

SUBORDER HAPLORRHINI
FAMILY CERCOPITHECIDAE

Young Females develop periodic (usually monthly) sexual swellings around the vulva, perineum and anus at the time of ovulation. They produce a single baby, rarely twins. Males fight for the right to mate with receptive females.

Habitat They range across the wooded savanna, grasslands, semi-deserts and rocky areas of Ethiopia, Somalia, Saudi Arabia and Yemen.

Golden Snub-nosed Monkey

The colobids and cercopithecids have been evolving separately since the middle Miocene era, about 12 mya. Colobid monkeys lack cheek pouches. They have large stomachs, divided into two chambers—a specialization developed to digest their almost exclusively vegetarian diet.

Close relatives Some authorities now classify this species in a separate family, Colobidae, with their relatives: langurs, other snub-nosed species, colobids and the proboscis monkey.

Characteristics Female snub-nosed monkeys weigh up to 22 pounds (10 kg), males are one-third as heavy again. Their leaf-like noses are markedly upturned. They have four long fingers on each hand and very short thumbs, a typical feature of colobus monkeys.

RHINOPITHECUS ROXELLANA
The beautiful shaggy coat of these rare monkeys was said to protect the wearer from rheumatism. Fortunately, only members of the Chinese Imperial family were permitted to wear it. The species now faces deforestation.

CLASSIFICATION

SUBORDER HAPLORRHINI
FAMILY CERCOPITHECIDAE

Food It feeds on leaves, fruit, buds, bark, pine-cone seeds and bamboo shoots with supplementary insects, birds and eggs.

Young Breeding tends to coincide with peak availability of food. Single, occasionally twin, infants are born.

Habitat Golden snub-nosed monkeys inhabit the cool mountain forests of southern China, Tibet and eastern India.

Dusky Leaf Monkey

Langurs, or leaf monkeys, fall into four distinct groups based on the fur color of the newborn young. They are the Asian representatives of the colobid monkeys, differing from their African relatives in having thumbs—but their leaping skills are quite comparable.

Close relatives The dusky leaf monkey is one of 15 species of leaf monkeys.

Characteristics The dusky leaf monkey has a head–body length of 17–27 inches (43–68 cm), with a slightly longer tail at 22–34 inches (57–86 cm). As with all colobid monkeys, leaf monkeys have large stomachs divided into two chambers. These fermentation chambers break down cellulose to make extra nutrients available from the vegetarian diet.

Suborder Haplorrhini
Family Cercopithecidae

Food Leaves are the major component of the leaf monkey's diet.

Young When a new troop male succeeds to a harem, every 27 months or so, he tries to kill any unweaned babies, behavior quite common in primates. A female that loses a suckling young stops lactating and is ready to mate almost immediately with the new troop leader.

Habitat The dusky leaf monkey inhabits forests, scrub plantations and gardens from north-eastern India to the Malay Peninsula.

SEMNOPITHECUS OBSCURUS
The dusky leaf monkey lives in extended families of 15–20 animals. The females are devoted mothers and the young, which grow darker fur as they grow older, can count on protection from adult group members until they are about two years old.

LAR GIBBON

Gibbons are the smallest of the apes. The apes differ from their sister-group, Old World monkeys, in having no tail; long arms; and mobile shoulders and wrists. Gibbons have extraordinarily long arms and move easily and rapidly through the trees by brachiation. The best-known species is the white-handed or lar gibbon. It depends upon ever-shrinking forests for survival.

Close relatives There are nine species of gibbons, and while grouped with the hominoids, they are only distantly related to the orang-utan, gorilla, chimpanzee and human. They are sometimes classifed in a separate family, Hylobatidae.

Characteristics Pairs live together in tree-top territories, with up to four offspring. The male calls in quivering hoots, the female in longer notes. The pair often "sing" in the early morning emphasizing their territorial claim and their bonding. These small apes (about 15 pounds or 7 kg) have remarkably dense fur: two to three times that of most monkeys. They may be buff or dark brown-black in either sex.

HYLOBATES LAR

Like all gibbon species, the lar gibbon is highly arboreal and moves through the trees by brachiation, that is, swinging its long arms from branch to branch with alternate hand-holds. They walk erect on the ground and along large branches.

Food Lar gibbons eat fruit, young leaves, shoots, buds and flowers and occasional insects.

Young Baby lar gibbons are born every two to four years and are suckled for at least two years. The adult male carries and baby-sits the infant at least beyond its first year.

Habitat Lar gibbons can be found in northern Sumatra, the Malay Peninsula (where they are still quite common), Thailand, parts of Myanmar (Burma) and Yunnan.

CLASSIFICATION

SUBORDER HAPLORRHINI
FAMILY PONGIDAE

Common Chimpanzee

The skills of these intelligent creatures include the ability to press buttons on keyboards to communicate words. In the wild, about 13 different chimp calls have been identified, from soft grunts to far-carrying shrieks and roars. Like the gorilla, it is a specialized knuckle-walker, is black-skinned, and has large brow ridges. However, it appears to be more closely related to humans.

Close relatives There are two species in the genus, *Pan*.

Characteristics Chimpanzees have similar sensory abilities to humans but are probably better able to distinguish smells. They forage in large communities of 20–100 or more, made up of smaller family subgroups. They spend about half their time on the ground. In the trees, they knuckle-walk along branches or brachiate beneath them.

Food Chimpanzees mix a little vertebrate prey in with their otherwise vegetable diet and form groups to hunt monkeys.

Young Newborn chimpanzees are helpless. Riding piggyback does not begin until about five months and lasts, as does lactation, for several years. There is a four-year interval between the births of usually one young.

Habitat Chimpanzees range across the forests, woodlands or mixed savannas of central and west Africa.

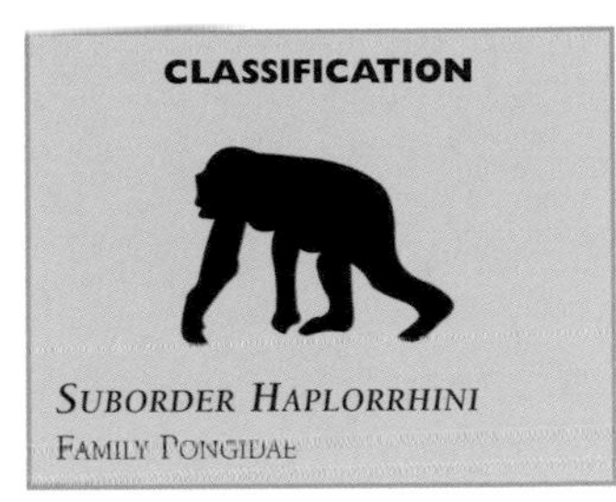

PAN TROGLODYTES

Chimps spend most of the daylight hours on the ground, associating in small parties, either roaming widely, searching for fruiting trees, or in more sedentary nursery groups of mothers and infants. They use tools for food gathering: cracking nuts open with rocks, sweeping ants or termites up with pieces of wood, or enlarging holes in trees with sticks to get at honeycomb.

ORANG-UTAN

Orang-utan is Malaysian for "man of the woods" or "wild person." These animals are solitary forest dwellers. Males wander alone; females are accompanied by a single offspring, if they have one. The long call of the mature male orang-utan is probably used to delineate territory and to attract sexually receptive females. They appear the least humanoid of the great apes, but are remarkably intelligent and adaptable creatures.

PONGO PYGMAEUS
Orang-utans use their arms and hindlimbs to move slowly through the trees. They do not brachiate, like gibbons and chimps. Young ones learn early to use their increasing body weight to bend trees in the desired direction of travel.

Close relatives There are two subspecies, the Bornean and Sumatran orangs, differentiated by their hair color and length, and the shape of the cheek flanges. The gorilla, chimpanzee and human are more closely related to each other than they are to the orang-utan.

Characteristics Orang-utans are truly arboreal apes and only occasionally come down from trees. At night, individuals construct sturdy sleeping platforms securely in the crown of a tree. When the orang lies in the nest, it makes a cover by pulling down other vegetation. Generally, a fresh nest is made each night. Fully mature males weigh between 176 and 200 pounds (80–91 kg) and continue growing until 13–15 years old.

THE RED APE
Sparsely covered in very long maroon to gingery colored hair, through which the rough, blue-gray skin can be seen, orang-utans are somewhat untidy in appearance. They have powerful, hook-shaped hands and feet—well developed for grasping. The great length of their arms makes their legs look short by comparison.

Food Orang-utans largely depend on fruit for nourishment but have been recorded taking more than 300 different foods from the forest, including fungi, bark, insects, birds' eggs, nestlings, baby squirrels, mineral-rich soil and honey. Tropical fruit trees at their disposal include durians, rambutans, jackfruits, mangosteens, mangos and figs. They drink water trapped in tree holes, dipping in a hand and sucking the drops from their hairy wrist. These intelligent animals remember the locations of their favorite fruiting trees over a wide area.

Young Orang-utans are sexually mature at about seven years. Females have an active reproductive life of about 20 years, during which time they give birth to three or four single offspring, which they raise on their own. Gestation averages 245 days.

Habitat Orang-utans are now restricted to northern Sumatra and most of lowland Borneo. Deforestation is threatening their home. Logging, the land hunger of an increasing human population, and deliberately lit fires are devastating the orang habitat. Pusat Rehabilitasi Mawas in north Sumatra and other centers elsewhere in Sumatra and in Kalimantan preserve wild orang-utans (named "mawas" in Sumatra) from slaughter or capture and rehabilitate confiscated specimens.

MAN OF THE WOODS
Orang-utans have large brains. The exact ancestral lineage of the contemporary species has not yet been traced, but these animals clearly have links with *Sivapithecus* fossil apes dating back to the Miocene era.

Gorilla

The largest living primate is generally considered the most intelligent land animal apart from humans. It is capable of learning words in human sign language. Older males are called silverbacks because they develop silvery gray hair on their backs when they are fully mature after about 10 years of age. Fossils and biochemical data indicate that chimpanzees and gorillas are more closely related to humans than to orang-utans.

Close relatives There are three subspecies or races of *Gorilla*.

CLASSIFICATION

Suborder Haplorrhini
Family Pongidae

Characteristics These bulky animals are knuckle walkers, that is, they move on all fours on the soles of their hindlimbs, but pivot their flexed hands on the ground with the weight on the middle joint of the fingers. They are very largely terrestrial and make simple nests at night. The average adult male weighs 385 pounds (175 kg) and stands 61 inches (156 cm).

Food Gorillas feed on vegetation.

Young A single baby is born after more than nine months' gestation, and clings to its mother, until three months old. It suckles for 12–18 months and stays with its mother for about three years.

Habitat Gorillas inhabit rainforest in the lowlands and mountains of Central Africa.

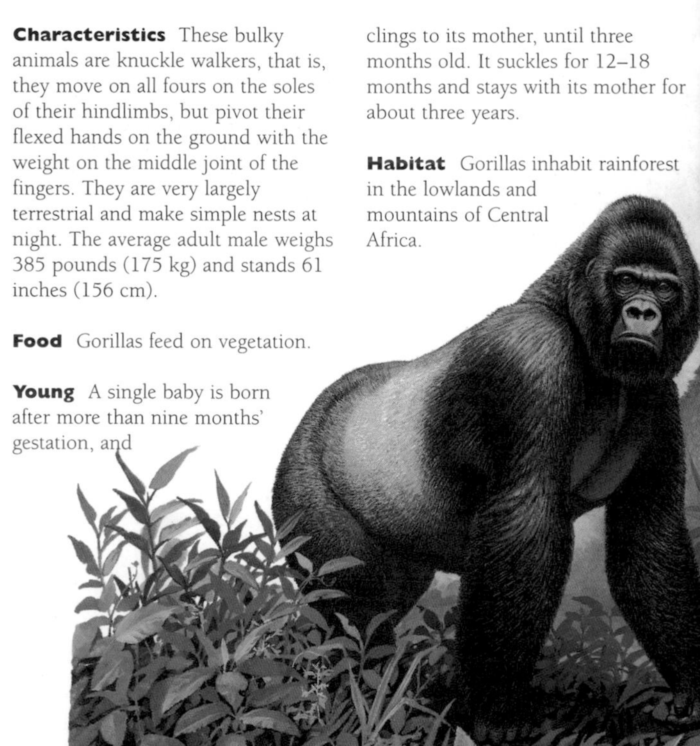

GORILLA GORILLA
Gorillas live in troops consisting of one silverback male (sometimes more), a few blackback males, and several females and young. The troop wanders over a large home range of 4–8 square miles (10–20 sq km), overlapping other troop ranges, with which relations are normally peaceful. Young gorillas are more arboreal than the heavier adults and swing themselves through the trees by brachiation.

GORILLAS AT RISK

Gorilla numbers are shrinking, primarily as a result of habitat loss. In Virunga Volcanos National Park in Rwanda, a sizeable section has been lost to grow pyrethrum, and loggers are constantly nibbling their way up the slopes. Even though African laws prohibit killing gorillas, enforcement is usually lacking. Poachers sell their heads and hands as souvenirs and they are also killed for food.

Most endangered The mountain gorilla is the most endangered of the three gorilla subspecies. Only 350–500 now survive. The greatest hope of saving them is through tourism. This undertaking is however, being disadvantaged by the struggles between rebels and the Rwandan government.

Zoo attraction Zoo administrators are well aware of how highly gorillas rate with the public and, although the trade has recently decreased, high prices are still paid for young specimens. Unfortunately, for every live one reaching a zoo, at least two die on the way.

KINDS OF MAMMALS

CARNIVORES

The order Carnivora consists of a unique group of mammals that share a common evolutionary history. The first members of the Carnivora had adaptations that allowed them to eat meat more efficiently than competing mammals. Not all Carnivora are meat-eaters; nor are all carnivores members of this order. The single most important feature that unites the order is a unique modification of the teeth for eating meat. They also evolved other modifications: acute eyesight, hearing, smell; speed and dexterity; the ability to kill and digest other animals; and greater intelligence than their prey. Carnivores are the only order of mammals that have ocean-going, terrestrial, arboreal and semi-fossorial species.

KINDS OF CARNIVORES

Carnivores have historically been seen as vicious predators, high on the food chain, even competing with humans. The 11 families, however, include animals that live entirely on plants, insects, large mammals and even fruit. Still, they are some of the most formidable of all mammals.

weasel

civet

hyena

Meat-eaters? Carnivory is a food habit of some, but not all, mammals in the order Carnivora. Cats eat meat as the principal part of their diet, but the giant panda feeds exclusively on bamboo. Other species specialize in eating fruit, insects, worms, nuts, berries, fish, crustaceans and seeds. Consequently, many members of the Carnivora are more appropriately described as opportunistic omnivores: they eat a wide variety of plant and animal foods, as they become available.

Diversity The smallest carnivore is the least weasel, at a mere 3 ounces (100 g); the largest is the southern elephant seal, weighing 5,300 pounds (2.4 tonnes). Within the group are the fastest land animals, the cheetahs and some of the slowest, the earless seals, better adapted to sea than land.

Carnivore distribution

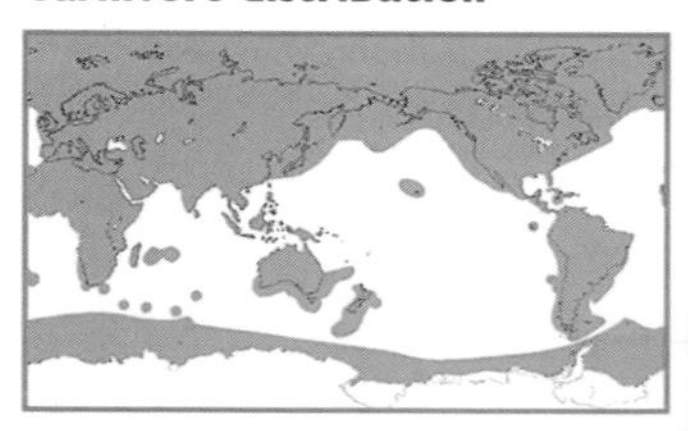

Carnivores are found on every continent in most habitat types: the oceans, the Arctic, the shores and pack-ice of Antarctica, tropical rainforests, prairies, temperate forest, deserts, mountains.

SPECIALIZED TEETH

Sharp-edged molars Carnivora are defined by the structure of their teeth. Two pairs of sharp-edged molars, called the carnassial teeth, are extremely efficient shearing edges. They are the fourth upper premolar, and the first lower molar. They are modified to form two vertical, sharp cutting surfaces, which slide against each other like the blades of a pair of scissors.

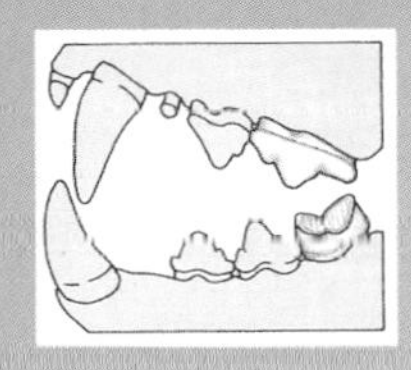

Modifications to the shear Carnivores that have evolved into non-meat eaters have modified the carnassial pair to suit their diet. Plant and fruit eaters have lost most of the vertical shear surface, and increased the horizontal crushing surface. Insect-eaters have smaller teeth with small shearing surfaces to pierce the exoskeleton of their prey.

Families The relationships among carnivore families are based upon the structure of the ear and sensitivity to various frequencies. They all have a highly developed ear region, which is attuned to sounds made by their particular prey.

Orders Carnivora are divided into two suborders: the dog-like carnivores (Caniformia); and the cat-like carnivores (Feliformia). They evolved from separate ancestors during the Eocene. Caniforms are mostly terrestrial, although the sealions and seals have become exclusively aquatic. The suborder includes the dogs, foxes and jackals; bears and pandas; raccoons and coatis; weasels, otters, skunks and badgers; sealions and fur seals; walruses; and true seals. Feliforms are terrestrial; only the otter civet and marsh mongoose are semi-aquatic. Included in the suborder are civets and genets; mongooses; hyenas and aardwolves; and lions, tigers and cats.

Maned Wolf

The dog-like carnivores are dominated by the canids. They are terrestrial carnivores, with long legs, bushy tails and slender bodies. They rely on hearing, smell and speed to locate prey. In Brazil, the excreta and body parts of maned wolves are said to have therapeutic qualities.

Close relatives The maned wolf is not a true wolf, but grouped as one of the three kinds of South American canids. It is most like a long-legged red fox.

Characteristics This is a large animal weighing about 50 pounds (22.5 kg) with extremely long legs for its body mass, clad in black "stockings." The adult maned wolf has a distinctive white-tipped tail, white patches at the throat and an erectile mane around the neck and across the shoulders.

Food Maned wolves devour fresh meat when they can catch it and fruits are about 50 percent of their diet.

Young Litters of up to five young are born once a year after a gestation period of about 65 days.

Habitat Found in remote areas in grassland and round the edges of swamps in South America.

CHRYSOCYON BRACHYURUS
The maned wolf is classified as vulnerable. Individual animals hunt mainly at night; their partiality for farm poultry and lambs has resulted in trapping or shooting. These free-ranging animals, whose stilt-like legs are perfectly adapted for moving through tall prairie grass, are highly adaptable and opportunistic feeders.

CLASSIFICATION

Suborder Caniformia
Family Canidae

Red Wolf

Red wolves, which once ranged freely throughout south-eastern USA, are extremely rare. Indeed, they may even already be extinct in their pure form, as they have been relentlessly hunted by humans since the late nineteenth century. They have also interbred easily with coyotes and probably now only exist as hybrids.

Close relatives Red wolves are related to true wolves and coyotes, the most social and carnivorous members of the canid family.

Characteristics Red wolves grow to a length of 41–49 inches (105–125 cm) excluding their tails, which add another 13–17 inches (33–43 cm). In its pure form, the red wolf is a cinnamon or tawny color with gray, black and white highlights.

Food Wolves hunt in packs, feeding on small mammals. They are intrepid and opportunistic scavengers.

Young It is unlikely that litters of pure red wolves are born in the wild.

Habitat If any red wolves still exist in the wild, they would be found on the coastal plains and in forests of central south-eastern USA.

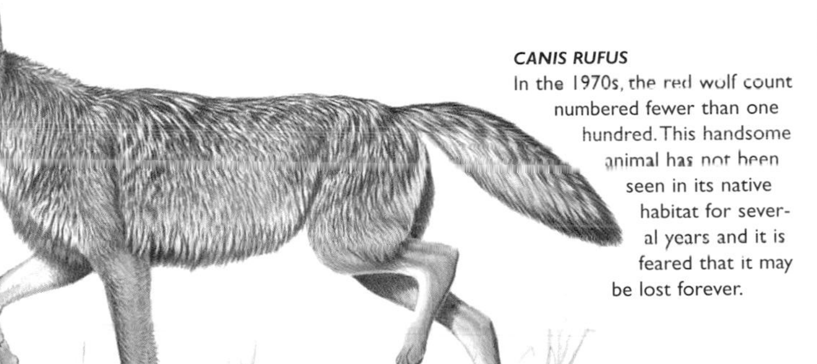

CANIS RUFUS
In the 1970s, the red wolf count numbered fewer than one hundred. This handsome animal has not been seen in its native habitat for several years and it is feared that it may be lost forever.

Gray Wolf

Also called the timber or white wolf, this animal was once the most widespread mammal, apart from humans, outside the tropics. There are probably two major reasons for its declining numbers: habitat destruction and the determined effort of country people to wipe it out. It is an intelligent and efficient predator that will take advantage of any feeding opportunity.

Close relatives Gray wolves are related to other members of the wild dog family, in particular, to the very rare red wolf and the much more plentiful coyote.

Characteristics Gray wolves are the largest of the dog family. In a typical dog-like way, they mark their territory with strongly scented urine. Like other Canidae, wolves will eat dead meat and have certainly consumed human corpses. They hunt in packs and have been recorded killing as many as 200 sheep in a night in Italy.

Food Pack members hunt together. Co-operative tactics allow them to pursue and fell animals that weigh up to 10 times more than they do, such as moose, deer, caribou and wild horses. Smaller mammals including hares, beaver and mice are summer fare. Their hunting

CLASSIFICATION

SUBORDER CANIFORMIA
FAMILY CANIDAE

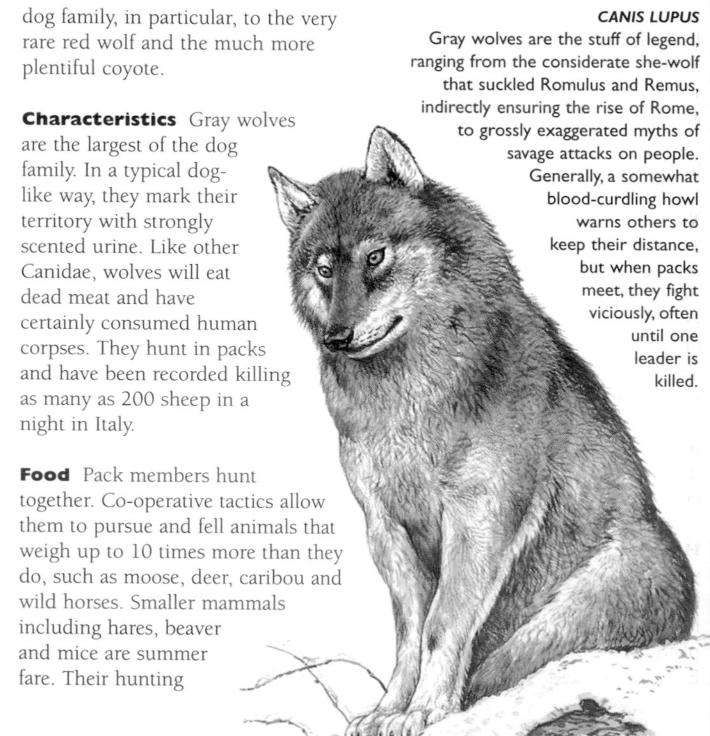

CANIS LUPUS
Gray wolves are the stuff of legend, ranging from the considerate she-wolf that suckled Romulus and Remus, indirectly ensuring the rise of Rome, to grossly exaggerated myths of savage attacks on people. Generally, a somewhat blood-curdling howl warns others to keep their distance, but when packs meet, they fight viciously, often until one leader is killed.

success rate is quite low, usually less than 10 percent, and they often have to go for long periods without eating. Gray wolves also eat fish and crabs, some plant material and human food scraps.

Young Gray wolves are sexually mature at around two years old and mate for life. Females bear litters of between four and seven blind, helpless young. These are cared for by their parents and other helpers in the group until they are strong enough to run with the pack.

Habitat The presence of gray wolves is much reduced but they are still found in a wide variety of habitats from mountains to desert plains in eastern Europe, the Middle East, North America and Asia.

HOWLING CHORUS

Wolves are social species and packs have a well-developed hierarchy centering on the dominant breeding pair. Co-operative hunting by 6–12 individuals allows them to specialize in very large prey. They communicate with body language, facial expressions and howls. The howling chorus can be heard for about 6 miles (10 km) and tells other wolves to keep away.

TAIL COMMUNICATION

No problems! The tail is in a relaxed position with the fur slightly fluffed when the wolf is eating or just comfortably standing.

Don't worry! A dropped tail with flattened fur and curled-back tip says "I'm no threat;" a stance assumed to beg a dominant colleague for food.

I'm frightened! The tail tucked between the legs touching the belly signals fear. A wolf that has just lost a fight holds its tail like this.

Watch out! A tail held straight out behind is a message of hostility. This animal is angry and plans to attack immediately.

Coyote

Coyotes are shrewd and highly adaptable; they have survived trap-ping, poisoning and other attempts to exterminate them since humans first domesticated animals hundreds of years ago. In North American folklore, the coyote is the chief trickster, witty and entertaining.

Close relatives There are more than 12 subspecies.

Characteristics Coyotes are light-boned, and have longer, narrower jaws and smaller ears and feet than their wolf relatives.

Food Coyotes can and do eat almost anything: carrion, freshly killed prey, insects or fruit. Lone coyotes will tackle small animals such as ground squirrels, mice and rabbits, but they gang up for larger meals on the hoof. Coyotes also catch fish and frogs.

CLASSIFICATION

SUBORDER CANIFORMIA
FAMILY CANIDAE

CANIS LATRANS
Coyotes communicate with one another through a unique repertoire of sounds: high-pitched yips, yelps, barks and wails. In the main, they are pack animals: a typical hierarchical group includes closely related adults, yearlings and young.

Young Pairs mate for life with dog-like devotion. Litters contain between five to 10 young and the male feeds the nursing female.

Habitat Coyotes, or prairie wolves, occupy a wide range of open country and forest, and are increasing their range from North Alaska to Costa Rica, throughout Mexico, the USA and Canada.

DINGO

The dingo is probably descended from the Indian wolf *Canis lupus pallipes*. Aborigines almost certainly brought dingos to mainland Australia in post-Pleistocene times and they have lived there, unchanged, for at least 8,000 years.

Close relatives Some authorities regard dingos as subspecies of domestic dogs and tag them *C. familiaris dingo*.

Characteristics Dingos live in packs and may hunt alone or co-operatively. In the wild they yelp or howl, but when domesticated may bark like tame dogs. The animal is stoutly built, standing about 24 inches (60 cm) tall at the shoulder, with short soft fur.

CLASSIFICATION

SUBORDER CANIFORMIA
FAMILY CANIDAE

Food Dingos run down kangaroos and feral pigs. They also eat grass-hoppers, lizards, the introduced rabbit and sheep.

Young Dingos readily interbreed with feral and domesticated dogs. They usually bear litters of four to five pups after a gestation period of about 63 days. Bitches suckle their pups for about two months.

Habitat Dingos range across most of the sandy deserts and wet and dry sclerophyll forests of mainland Australia, and similar forms inhabit many South-East Asian islands.

CANIS DINGO
Wild dingos are bold and suspicious, and packs may wreak havoc in a flock of sheep or prey upon poultry. Dingo-proof fencing has kept them out of some grazing territory. Dingos bred in captivity become affectionate pets.

African Hunting Dog

Co-operative hunting by groups of six to 12 hunting dogs, and their ability to sustain high speeds for considerable distances enable them to specialize in large prey.

Close relatives They are the sole members of their genus, but are most closely related to the true dogs and wolves.

Characteristics These animals have short muzzles and strong shearing teeth. The social arrangement of groups is unusual in that the males stay with the natal pack and females leave to try to breed elsewhere. Packs indulge in greeting rituals at least once a day, squeaking and rubbing muzzles with each other.

Food Least typical of canids, hunting dogs are entirely carnivorous.

Young Litters range from two to sixteen.

Habitat South of the Sahara desert to the Transvaal.

LYCAON PICTUS

The blotched pattern of the African hunting dog is unique among canids. These animals spend the greater part of the day resting and grooming in the shade. They hunt in the early morning, at dusk, or on bright moonlit nights.

AGGRESSION

Packs of African hunting dogs maintain friendly relations for most of the time, but females fight aggressively when competing to breed. They are able to run at 37 mph (60 km/h) for 3 miles (5 km) or more.

Bush Dog

This species is listed as vulnerable by the World Conservation Union. The bush dog is elusive in the wild and difficult to study, so that few specific details are available on its breeding habits and social life.

SPEOTHOS VENATICUS

The compact bush dog is atypical in appearance compared with other canids. It has short legs, a short snout and small ears. It lies up during the day in holes, and forages at night. Adults maintain contact with one another by whining.

Close relatives

Its relatives are the South American foxes and maned wolf.

Characteristics Like many other canid species, male bush dogs mark their territory with secretions from the anal glands as they cock their legs to urinate. Females back up to trees to urinate against the trunk from a "handstand" position. Bush dogs have sturdy teeth arranged in a formula unlike any other South American canid.

Food These carnivorous dogs swim and dive with agility, pursuing amphibious prey into the water. Packs hunting co-operatively can bring down large animals.

Young Litters of bush dogs are thought to number four to five pups. Males feed their nursing mates and both parents tend the young.

Habitat Bush dogs are forest and savanna dwellers, distributed through Central and South America.

CLASSIFICATION

Suborder Caniformia

Family Canidae

Gray Fox

The gray fox is the only vulpine species that habitually climbs trees and is also known as the tree fox. It is sometimes classified in the genus Urocyon, with the island gray fox from Santa Barbara Island, California, partly because of its longer skin gland near the base of the tail, as well as its climbing habit. The skin of all canids is almost completely without sweat glands, but there is a dorsal scent gland at the base of the tail.

Close relatives Twenty-one species of foxes are divided into four (or five) genera, with 12 species in the genus *Vulpes*.

Characteristics The canine teeth of the gray fox are shorter than the average for other foxes. The typically foxy face is shaped by the "brows" formed by the frontal bones above and between the eyes being slightly indented in *Vulpes*, rather than convex as in the genus *Canis*.

CLASSIFICATION

Suborder Caniformia
Family Canidae

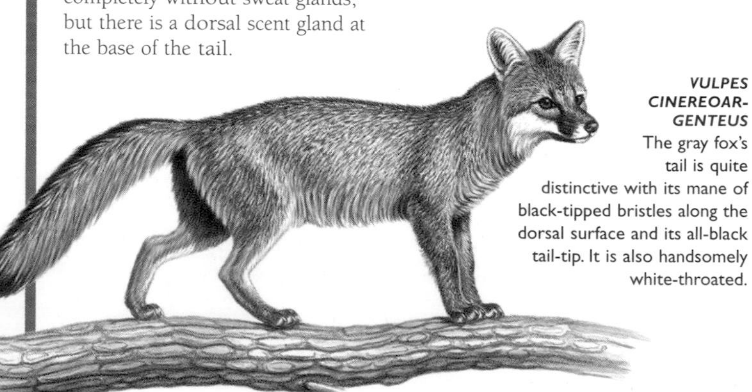

VULPES CINEREOARGENTEUS
The gray fox's tail is quite distinctive with its mane of black-tipped bristles along the dorsal surface and its all-black tail-tip. It is also handsomely white-throated.

Food Gray foxes pounce on rodents, birds and snakes with the characteristic leap of vulpine species. They also take birds' eggs when they come across them.

Young Like other foxes, gray foxes have annual litters of between one and six young.

Habitat Gray foxes inhabit the prairies of central USA south to Venezuela and north to Ontario.

RED FOX

In literature, the red fox is portrayed as wily and amoral. They may have a justifiable reputation for cunning, for they are remarkably resilient to human persecution. Millions have been slaughtered for their fur, to control the viral disease, rabies, to provide sport, or because they are feared around farmyards.

Close relatives Red foxes share the genus *Vulpes* with 11 other species.

Characteristics Typically active at night, red foxes annex the abandoned burrows of other animals or dig their own to use as daytime resting-places. They "mouse leap" onto prey, springing about 3 feet (1 m) off the ground and diving, front paws first, to squash their victims.

Food Red foxes prey largely on rodents. They also hunt rabbits, hares, birds, insects and invertebrates and frequently cache their prey for later consumption. Fall fruit and berries supplement their diet and they scavenge around refuse dumps in urban areas.

Young Red foxes breed once a year, litter sizes varying between four and eight young. Vixens have six teats. Birth dens have been located in hollow trees, long grass or under houses, though a burrow is preferred. Communal denning has been observed.

Habitat Red foxes are the most widely dispersed of all the carnivores, turning up in woodland, open country and even in towns in Canada, the USA (except for Florida and the Rockies), Europe (except for Iceland) and through Asia to Japan and Indo-China. They are feral in mainland Australia having been introduced there from Europe in 1871 for sporting purposes.

VULPES VULPES
The male red fox brings food to the family until the female is able to leave her cubs for short periods or take them out to learn the rudiments of foraging. Fox cubs indulge in prolonged bouts of play, good practice for hunting.

ARCTIC FOX

There are two color forms of Arctic fox in summer: brown, and steel-gray or blue. All Arctic foxes have a white winter coat, which assists them to hunt in snow.

Close relatives Arctic foxes are in a genus of their own with nine subspecies.

Characteristics Arctic foxes live in groups, generally containing a male and several vixens. Solitary night hunting is the norm; hunting in packs for small prey has no advantages. They weigh about 6–11 pounds (3–5 kg).

Food The main food of the Arctic fox is lemmings and even though other mammals (as carrion) and birds are occasionally eaten, the success of the fox and its young is determined by the lemming population cycle.

Young Vixens bear a litter of 4–17 young in May or June after a gestation period of 51–57 days. The average litter of 11 pups is much higher than for temperate zone red and gray foxes. Female Arctic foxes do not breed until at least two years old. In good lemming years, the survival rate of fox pups is higher. These characteristics are adaptations to living in a closely linked relationship with a prey species whose numbers fluctuate widely.

ALOPEX LAGOPUS

The thick coats of Arctic foxes are 70 percent warm underfur. This, together with their metabolism, enables them to withstand very cold conditions. They only begin to shiver when the temperature falls to –58°F (–50°C).

Habitat This species is distributed across tundra and open woodland in the Arctic regions of Europe, Asia and North America.

CLASSIFICATION

SUBORDER CANIFORMIA
FAMILY CANIDAE

Kit Fox

Kit foxes are solitary hunters and are strictly carnivorous. They are shy and uncommon, and unlike many other foxes, their numbers do not fluctuate, but remain remarkably stable over time.

Close relatives Kit foxes are one of the 12 species of vulpine foxes. This species is sometimes classified as a subspecies of the swift or plains kit fox (*Vulpes velox*), which occupies the same habitat (grasslands of central USA) and is slightly larger with smaller ears.

Characteristics The adults have pale fur and large ears. Their reputation for speed is based on the illusion created by their short legs and rapid gait.

Food Kit foxes primarily prey on rodents, rabbits and insects, but like other fox species, they are opportunistic feeders.

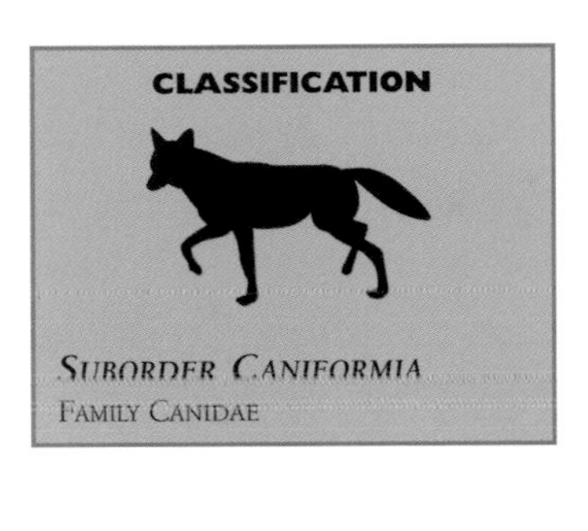

Young Vixens produce a litter of young once a year. While rearing the cubs, the female rarely leaves the den, the male providing all the prey for her and the young.

Habitat This species is a prairie-dweller of the desert and steppelands in south-western USA and north-western Mexico.

Vulpes macrotis
The typical long tail of this fox is slightly bushier than many other species. Kit foxes are solitary hunters, and cover large distances in search of prey. They use "mouse-leaping" and stealthy hunting techniques.

Bat-eared Fox

This is one of the most peculiar of the canids, and the only one that is primarily insectivorous. Not only do the large ears of this species assist the animals to hear acutely, they also dissipate heat—an important aid to coping with the climate of their arid environment.

Close relatives Also known as Delandi's foxes, they are in a genus of their own.

Characteristics This species is unique among carnivores for its dentition. Its teeth are relatively small but it has up to eight extra molars, which provide greater chewing surface.

Food Bat-eared foxes favor insects for 80 percent of their diet. They eat colonies of harvester termites and dung beetles, both found where herds of large ungulates feed. Fruit, scorpions and occasionally a mammal or bird supplement the menu.

Young Pairs of bat-eared foxes usually breed in self-dug dens where they raise between two and five pups. Two to three breeding dens are sometimes clustered together. Juveniles are full-grown at four months old.

Habitat This species hangs around hoofed mammals, such as zebras, buffalo and wildebeest, on the African grasslands. The first distinct population is found from south Zambia to South Africa; the second population from Ethiopia to Tanzania.

OCTOCYON MEGALOTIS
Aptly named, bat-eared foxes use their unusually large auditory organs to detect the sound of dung beetle larvae gnawing their way out of dungballs. They occasionally hold their tails aloft in an inverted U-shaped position when communicating with one another.

Raccoon Dog

Raccoon dogs are so named because their black face masks are very like the North American raccoon. They are unique among canids for not barking, for hibernating in winter and living in the dense undergrowth of forests.

Close relatives Raccoon dogs are the only species in their genus.

Characteristics Primarily nocturnal, the raccoon dog is an excellent swimmer. It lives alone or in family groups of five to six animals. Raccoon dogs are commercially valuable in the fur industry and have been introduced in many areas.

Food Raccoon dogs are omnivorous and vary their diet according to seasonal availability of fruits, insects and other invertebrates. They are especially fond of frogs and fish.

Young Females bear six to eight young after a gestation period of approximately two months. The pups are able to fend for themselves at about six months old.

Habitat The raccoon dog inhabits forest and rocky banks near lakes and rivers. It is found in eastern Siberia, north-east China, Japan, northern Indo-China and is an introduced species in east and central Europe.

NYCTEREUTES PROCYONOIDES
Raccoon dogs have thick soft body fur, which is commercially valuable. They are hunted not only for their skins but also as a pest because they carry rabies.

CLASSIFICATION

Suborder Caniformia
Family Canidae

Asian Black Bear

The ancestors of the seven species of modern bears first appeared in the Oligocene, from an earlier line shared by the canids. Asian bears are not always black; the color of their fur also ranges from reddish to dark brown.

Close relatives The Asian black bear is monotypic in its genus.

Characteristics These large beasts may be black or reddish to dark brown, with white markings on the snout and a white crescent on the chest. The Asian black bear is a good climber. It sometimes sleeps on tree platforms high above the ground that it builds from branches broken when it is feeding.

SELENARCTOS THIBETANUS

Asian black bears share a great similarity in body form with all the modern bears, except the panda. Their ancestor was a red panda-sized animal. Throughout evolution bears have increased in size, becoming the largest of the land-dwelling carnivores.

CLASSIFICATION

SUBORDER CANIFORMIA

FAMILY URSIDAE

Food This species eats fruit, nuts, insect larvae and ants. They will kill domesticated sheep, goats and cattle.

Young Two Asian black bear cubs form a litter; they are very small and blind at birth.

Habitat These bears range up to 11,800 feet (3,600 m) in the forests and brush of Afghanistan, China, Siberia, Japan, Korea, Taiwan, Hainan and South-East Asia.

AMERICAN BLACK BEAR

The American black bear may also be white, blue and brown. It is a slow-moving animal not much inclined to vigorous activity. Its territory occasionally overlaps with the much bigger brown bear—the larger animal will occasionally kill the smaller one.

URSUS AMERICANUS
American black bears climb trees with ease to look for ripe fruit or honeycomb or to take a nap. Frightened cubs will often shin up a trunk to avoid danger and wait out the threat there until their mothers come to fetch them down.

Close relatives American black bears look like a smaller form of grizzly (brown) bears to which they are closely related.

Characteristics American black bears grow slowly; their natural life-span is about 30 years and females may only produce six to eight cubs in a lifetime. They generally feed at dawn and dusk and may bed down in swamps during the day.

Food American black bears eat fruit, nuts, roots, honey, insects, rodents, small mammals, stranded fish, carrion and human refuse.

Young Twins are the norm, though litters of up to five cubs have been reported. Newborn cubs are the size of rats and without fur. Mothers produce young in alternate years during the denning season.

Habitat These bears inhabit wooded areas, swamps and mountains in Alaska, Canada and North America. They are also found in north Mexico.

CLASSIFICATION

SUBORDER CANIFORMIA
FAMILY URSIDAE

Brown Bear

Once plentiful land-dwelling mammals, brown bear numbers have been dwindling steadily.

Close relatives These are the polar and American black bears.

Characteristics Brown bears' claws extend to 4 inches (10 cm), they can travel at 30 mph (50 km/h) and are strong enough to carry the carcass of a bull moose. The winter torpor is not true hibernation since it does not slow heart and breathing rates drastically or lower blood pressure or temperature.

Food Tubers, berries, fish and carrion form the diet.

Young Females give birth to one to four cubs in their winter dens.

Habitat A forest and tundra dweller in remnant populations in the Northern Hemisphere.

URSUS ARCTOS

Brown bears assume the upright, totem-like stance when they want to intimidate or are simply surveying their surroundings. More usually they lumber about on all fours. Their eyesight is poor and their hearing indifferent, but the moist, prominent nose is finely tuned. Brown bears have few predators and are not designed to move quickly over great distances. The soles of their hindfeet are flat on the ground; stout limbs and a short back equip them for digging, fishing and climbing.

CLASSIFICATION

SUBORDER CANIFORMIA

FAMILY URSIDAE

POLAR BEAR

Despite being the largest of the terrestrial carnivores, polar bears are swift on their feet and can outrun a reindeer over a short distance. They are mostly solitary animals.

Close relatives Polar and brown bears have a common ancestor.

Characteristics The polar bear has a thick, water-repellent winter coat and an insulating fat layer. Only the nose and foot pads are not furred. Very large mature males of 8–10 years can weigh 1,760 pounds (800 kg). Mature females are about half this weight but become extremely fat when pregnant and occasionally exceed 1,100 pounds (500 kg).

Food They feed almost exclusively on marine mammals, especially seals.

Young Females become pregnant every three years, after four years old, and usually have twins.

Habitat Polar bears live on the ice floes and along the coasts of the Arctic Ocean to the southern limits of the floes.

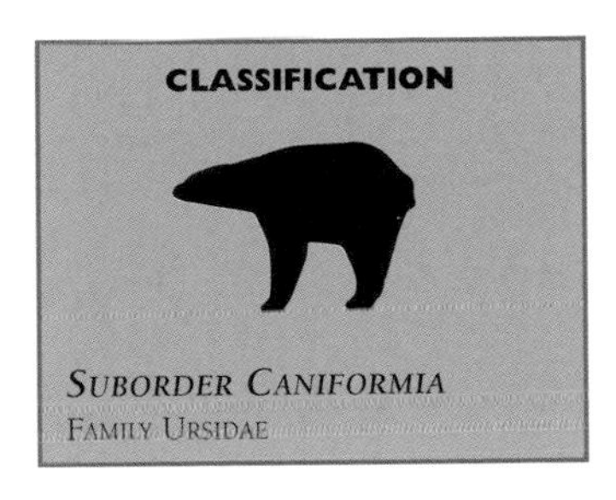

CLASSIFICATION

SUBORDER CANIFORMIA
FAMILY URSIDAE

URSUS MARITIMUS
Mothers and cubs share a family bond for about two-and-a half years. Cubs weigh 2 pounds (less than 1 kg) at birth. They stay snugly within the maternity den, a hollowed out space in wind driven snow, for three to four months suckling fat-rich milk. When occupied, the inside temperature of a den can be 68°F (20°C) warmer than it is outside. Males behave aggressively toward one another during the breeding season; they take no part in family life and occasionally kill cubs.

A POLAR BEAR'S YEAR

The polar bear is the largest terrestrial carnivore; the only truly carnivorous bear; and the only maritime bear. It is superbly adapted to life in its extremely harsh Arctic environment. In search of their primary prey, ringed seals and to a lesser degree bearded seals, they may travel on ice floes up to 50 miles (80 km) per day.

Late February to April Mother and cubs leave the den at a time when the mother can replenish her fat stores by feeding on ringed seal pups. Mother's keep their cubs for 2½ years before mating again.
March and April Each year, adult male polar bears walk many miles in search of breeding females, then fight for breeding rights.
April and May Ovulation is induced by mating frequently over a period of one to two weeks.
April to July Ringed seals, the polar bear's main prey, give birth in April, and their pups are weaned six weeks later. The nursing and recently weaned young seals available during this period can be up to 50 percent fat, and naive about predators.

August to November In late July, the sea ice melts and polar bears come ashore, living on their stored fat until freeze-up, when they are able to hunt seals once again.
September The fertilized egg, called a blastocyst, is implanted in the uterus.
Late October to November The pregnant female digs her maternity den in snowdrifts, usually on land within a few miles of the coast.
November to January The pregnant female gives birth and nurses her cubs inside the den, remaining there until the Spring.
December to February With the exception of pregnant females, polar bears are active throughout winter, using temporary dens and living on stored fat during periods of intense cold or storms.

April and May
April to July
September
August to
November
December to
February

Sun Bear

In Thailand, sun bears are known as dog bears as they are small and stocky. In Malaysia and Indonesia, these animals are called honey bears, emphasizing their liking for wild honey. Their most common name refers to the yellowish or white crescent-shaped mark on the chest.

Close relatives The species is monotypic in its genus.

Characteristics Sun bears are the smallest of the world's species of bears. Their fur is short and sleek, a possible adaptation to the lowland equatorial climate. They are nocturnal, and do not become dormant at any time of year.

Food Sun bears have a long tongue, which they use for licking up termites after they have broken open the mound, and for licking up honey from wild bees' hives. They also feed on insects and their larvae, fruit and coconut palms and hunt jungle fowl and small rodents.

HELARCTOS MALAYANUS
Sun bears are excellent tree climbers. Like Asian black bears they make platforms of broken branches above the ground on which they sleep or sunbathe during the day, stirring mainly at night to go foraging for food.

Young It is not known where sun bears give birth to their cubs or how they deal with newborn offspring. Females accompanied by one or two cubs have been observed.

Habitat Sun bears are found in the forested mountains and lowland of South-East Asia, including the islands of Sumatra and Borneo.

Sloth Bear

The common name for this bear is derived from the mistaken identity of the first specimen described by a European zoologist, George Shaw, who thought it was a sloth because it had long, curved claws and lacked the incisor, or front teeth.

Close relatives Sloth bears are a single species in their genus.

Characteristics Male sloth bears are bigger than females, weighing between 198 and 253 pounds (90–115 kg) with oversize specimens sometimes reaching 297 pounds (135 kg). These nocturnal bears have powerful front teeth, long snouts and tongues, typical of insect-loving mammals.

Food The sloth bear's claws are used to rip open termite mounds and the gap in the teeth forms a funnel with the tongue and lips through which the bear sucks up insects. They also feed on sugarcane, honey, grubs, eggs, fruit, flowers and carrion.

Young Sloth bears do not become dormant, but females den for seclusion and protection when they give birth. Cubs stay with their mothers for two to three years. It is thought that these bears mate for life.

MELURSUS URSINUS

Sloth bears have a long, shaggy coat, which is almost invariably black. Their muzzle is white to chestnut brown and complemented by a U- to Y-shaped mark on the chest.

Habitat The lowland forests of east India and Sri Lanka are home to this small bear, which is listed as a vulnerable species.

CLASSIFICATION

SUBORDER CANIFORMIA
FAMILY URSIDAE

GIANT PANDA

The *Shi Jing*, China's earliest collection of poetry, dating from 1000 BC, describes the panda as "like a tiger and like a bear." The earliest Chinese dictionary (200 BC) describes the animal as a black and white bamboo-eating leopard.

Close relatives The giant panda appears to be more closely related to the bears than its bamboo-eating namesake, the red panda.

Characteristics Pandas are unusual in that they have the stomach and intestines of a carnivore but subsist on a herbivorous diet. Giant pandas have specialized forepaws for handling bamboo with an enlargement of the wrist bone to function as a sixth digit.

Food The branches, stems and leaves of bamboo form 99 percent of the panda's daily food, but it will occasionally consume other plants, and rarely eat carrion.

CLASSIFICATION

SUBORDER CANIFORMIA
FAMILY URSIDAE

Young Male and female seldom meet except to mate. Females generally raise only a single cub. Newborn cubs are blind and nearly naked and squawk loudly.

Habitat Giant pandas live in the mountainous forest of central China. They have survived for millions of years but are now threatened by the destruction of their natural habitat.

AILUROPODA MELANOLEUCA
Adult giant pandas weighing 220 pounds (100 kg) must eat 26–33 pounds (12–15 kg) of bamboo leaves and stems or 50–84 pounds (23–38 kg) of bamboo shoots a day to maintain condition. They spend the three months of summer eating, becoming active at twilight.

Red Panda

Molecular studies have shown that the red panda holds a position between the bear and the raccoon families and some scientists suggest that the red panda might be a surviving member of an extinct group of carnivores. It is similar in size to its ancestors.

Close relatives The red panda is unique in its family.

Characteristics Red pandas tend to be nocturnal and move about alone using scent from their anal glands to mark their territory. They also defecate regularly at chosen sites, which may be another way of staking their claim to territory. Like the giant bamboo-eaters, red pandas also have an extra "thumb."

Food Red pandas eat on the ground, preferring bamboo shoots, grass, roots, fruit and acorns.

Young Red pandas give birth from mid-May to mid-July in hollow trees. Litters may contain one to four offspring but two are usual. The young are born blind and helpless, but fully furred. They are weaned at about five months and become sexually mature at 18–20 months.

Habitat They live in bamboo forests from Nepal to western Myanmar and in south-west China.

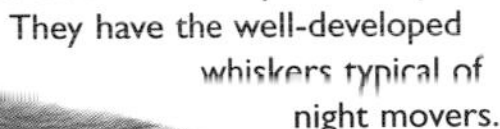

AILURUS FULGENS
The red panda's striped bushy tail and facial markings make it look much more like a raccoon than like its giant black and white relative. Red pandas climb trees during daylight hours and find secluded spots in which to curl up and sleep. They have the well-developed whiskers typical of night movers.

Common Raccoon

Members of the family Procyonidae only occur in the New World and are defined principally on morphological features of the skull and carnassial shear. The common raccoon is the best known of the group, familiar to most North Americans.

Close relatives The family comprises raccoons, olingos, coatis, kinkajous, ringtails and cacomistles, all linked to the dog–bear line of evolution. The genus *Procyon* contains six species.

PROCYON LOTOR

This animal's name originated from a North American word *aroughcan* or *arakan*, which means "he who scratches with his hands." Raccoons occur in many urban areas and frequently raid trash cans after dark. Many raccoons die on the roads and in the annual "coon hunting" between September and December.

Characteristics Common raccoons have highly dexterous forepaws. They have the habit of rubbing, feeling and dunking their food in water. They are nocturnal, solitary and found mainly in forested areas, where they live in dens in hollow trees or rock crevices.

Food Raccoons are night foragers. They often seek aquatic creatures such as frogs, crayfish and fish. The rest of their wide-ranging diet includes small rodents, birds, nuts, seeds, fruit, corn, and turtle and birds' eggs.

CLASSIFICATION

Suborder Caniformia

Family Procyonidae

Young Raccoons give birth in ground burrows in spring; litters contain between three and seven babies. The young's eyes are closed for three weeks. They venture out with mother at about two months and remain with her until fall.

Habitat Common raccoons are plentiful in diverse habitats across southern Canada, the USA and Central America, south to Panama. They are often found near water.

Ringtail

Ringtails came by their alternative common name, miner's cats, because early prospectors in the American West kept them as mousers and pets. The ringtail is one of the smallest and most carnivorous members of the raccoon family. It is unique in retaining the sharp cutting edge of the carnassial shear, while its rela-tives have blunt cusps.

Close relatives A closely related, larger species occurs from southern Mexico to Costa Rica.

Characteristics This predator is strictly nocturnal. It has dog-like teeth, semiretractile claws and larger ears than other procyonids. The head and body length is 12–15 inches (31–38 cm).

Food Ringtails prey heavily on lizards and small mammals up to the size of rabbits. They also eat large insects, nuts and fruit.

BASSARISCUS ASTUTUS
Ringtails are lithe, long-bodied animals, active at night. They are fairly common but are seldom seen because they hide during the daytime in secluded spots, probably in trees.

Young One to four young, rarely five, are born after a gestation of 51–54 days in May or June.

Habitat Ringtails inhabit dry areas such as rocky cliffs in western USA from Oregon and Colorado south to the Isthmus of Tehuantepec in Mexico.

CLASSIFICATION

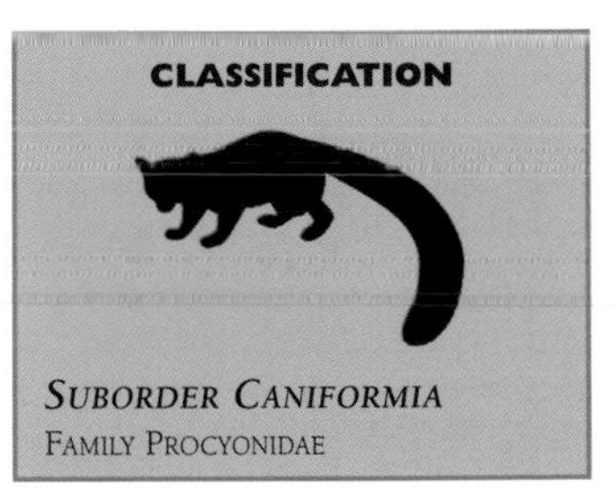

SUBORDER CANIFORMIA
FAMILY PROCYONIDAE

Coati

Adult female coatis and their offspring form foraging groups from four or five to 50 individuals. Males are solitary outside the mating period and are chased away from the group by females because they sometimes kill the juveniles.

Close relatives Coatis are closely related to raccoons and other procyonid species.

Characteristics Unlike other procyonids, coatis are mainly active in daytime. They have highly flexible snouts, which they are able to poke into holes and crevices in search of food. Among a single family, there may be dramatic differences in coloration: red, gray, black and brown.

Food Coatis are omnivorous. Their staple diet is mainly roots, insects, scorpions, spiders and other small ground-dwelling invertebrates. They also eat ripe fruit and unearth turtles' and lizards' eggs.

CLASSIFICATION

SUBORDER CANIFORMIA
FAMILY PROCYONIDAE

Young Coatis mate in trees. Females make platform nests for their young or use the crowns of large palm trees as natural nests. Litters contain three to five young, poorly developed at birth; they are nestbound for five to six weeks.

Habitat Coatis are distributed in woodland and lowland throughout the southern USA and Panama and most of South America, except Patagonia.

NASUA NASUA
The coati uses its long tail as a balancing rod when climbing quickly through trees. Foraging bands spread out over the forest floor, moving slowly along, digging and rooting with their forepaws and flexible snouts, and investigating all possible sources of food.

Kinkajou

The kinkajou is unusual in the Carnivora. It feeds primarily on fruit and is the only New World carnivore to have a prehensile tail. This skilled climber moves with speed and agility and lives almost entirely in the trees. It is no wonder that it is sometimes mistaken for a primate.

Close relatives The kinkajou (genus *Potos*) is strikingly similar in external appearance to the coatis (genus *Nasua*), though the latter lack a prehensile tail.

CLASSIFICATION

Suborder Caniformia
Family Procyonidae

POTOS FLAVUS
Kinkajous are fast-moving, extremely agile and travel constantly during the night. Groups will often feed in the same fruit trees that monkeys have used during the daytime.

Characteristics These animals are almost entirely arboreal. They are 16–24 inches (40–60 cm) long and weigh 3–5 pounds (1.5–2.5 kg). They have foreshortened muzzles, an elongate body form and uniformly colored short fur.

Food Besides fruit, kinkajous are very partial to sugary foods and use their exceptionally long extrudable tongues to probe blossoms for nectar and to gather honey.

Young Kinkajous are year-round breeders. They produce an annual single offspring. Young are able to hang by the tail at about eight weeks and are independent by four months old.

Habitat Distributed in tropical forests in east Mexico, through Central and South America to Brazil.

Sea Otter

Mustelids are the most successful evolutionary group of the Carnivora: diverse and numerous.They are distinguished by the musk gland near the anus and unique reproductive biology. The fur of sea otters is so highly valued that the animals were almost hunted to extinction by 1911.

ENHYDRA LUTRIS

The sea otter fetches its food from the ocean bed. It eats floating on its back, sometimes using a flat stone placed on its chest as a tool. It grips its prey in its forepaws and repeatedly smashes the hard shell of a mollusk against the rock until it breaks.The dye from sea urchins colors otter skeletons purple.

Close relatives There are 12 species of otters, grouped in a family with weasels, skunks and badgers.

Characteristics Sea otters are not insulated against the cold water by a layer of fat; they depend on the layer of air trapped in their dense coats for warmth and groom themselves constantly to keep the fur sleek. Sea otters have webbed feet and stiff whiskers, which they use as tactile sensors. Otters sleep on the surface of the water, entangled in kelp to prevent them from drifting. Sea otters, unlike most other mustelids, are gregarious, but males and females generally form separate groups.

CLASSIFICATION

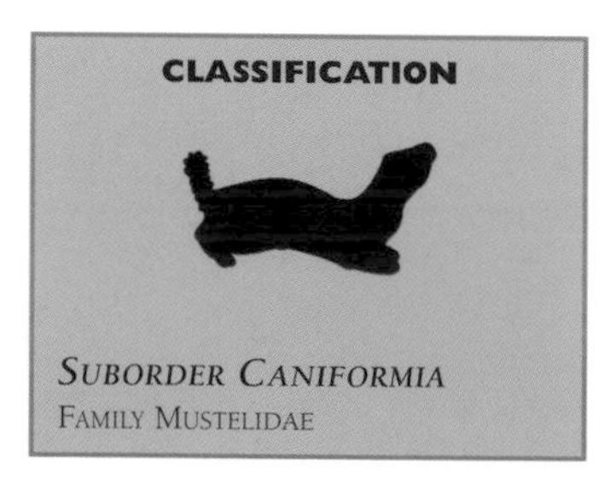

SUBORDER CANIFORMIA
FAMILY MUSTELIDAE

Food Sea otters feed on clams, sea urchins, mussels and abalone, using "tools" such as flat rocks to break open mollusk shells.

Young Otter pups are born fawn-colored with open eyes and a full set of milk teeth. Females bear a single offspring and nurse them in their chests as they swim on their backs.

Habitat This almost completely marine animal lives around the Kurile and Aleutian Islands and in the Gulf of Alaska; and in the Pacific Ocean along the coast of North America and former USSR.

Wolverine

In the sixteenth and seventeenth centuries, wolverine pelts were so valuable that the city of Turinsk in Siberia used this animal to represent its trade interests on its coat of arms. Wolverines are still shot and trapped in some places.

Close relatives Two subspecies of wolverines share a genus, and are grouped in a family with weasels and martens.

Characteristics Wolverines are ground-dwelling animals, but they can and do climb trees. They have large feet, which give them a speed advantage in soft snow when hoofed mammals are slowed down.

Food This heavily built and largely carnivorous animal is capable of killing animals larger than itself such as reindeer and caribou in winter. Summer prey includes small and medium-sized mammals and birds. As well as hunting, wolverines seek out dead animals. They also eat berries and plants.

Young Delayed implantation of the fertilized egg means that wolverines can mate and give birth at the most suitable times. Females usually dig a birth den in a deep snowdrift where one to four kits are born blind. They suckle for two months, remain with mother for up to two years, and are sexually mature at about four years.

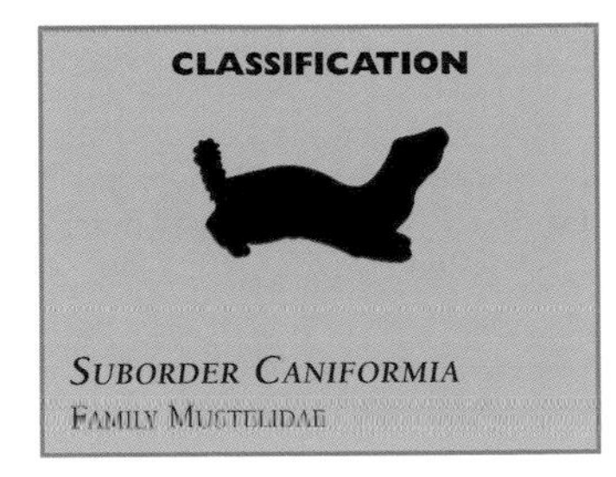

CLASSIFICATION

Suborder Caniformia

Family Mustelidae

Habitat Wolverines live in coniferous forests and Arctic and subarctic tundra in Scandinavia, Siberia, Alaska, Canada and western USA.

GULO GULO

Also known by their medieval name, glutton, wolverines are solitary hunters and more often the victor than the victim. However, they may be killed by packs of wolves, grizzly (brown) bears and pumas.

SKUNK DEFENSE

In skunks, the anal gland is modified to an organ capable of ejecting a fetid fluid over large distances with considerable accuracy. There is no mistaking the intentions of a skunk performing its elaborate ritual of intimidation.

Threats Most skunk species warn predators before they spray them by stamping their front feet, raising and fluffing up their tails, and walking with a stiff-legged gait. Skunks spray on all fours; the handstand position is a threatening posture.

Action They aim at the intruders' face and, apart from the noxious smell, the fluid causes intense irritation and even temporary blindness. In a favorable wind, the spray can travel for 13–23 feet (4–7 m), although accuracy is usually only guaranteed over a distance of $6\frac{1}{2}$ feet (2 m).

A skunk will confront its predator face-to-face; even an animal much larger than itself.

It raises its tail as a warning that attack will follow if the predator does not retreat.

The skunk then turns its back on the aggressor, and raises up on its forelegs, in a further threatening gesture preparing for attack.

The skunk ejects a stream of foul-smelling liquid, aiming toward the mouth and eyes of the victim.

Striped Skunk

The striped skunk is one of the better known animals in the mustelid family. It is notorious for the pungent smell of its anal gland secretions, which can be projected for a great distance. An effective defense against enemies, the odor temporarily stops the victim breathing.

Close relatives Members of the same genus are the hooded skunks.

Characteristics Striped skunks are nocturnal. Their tails are covered with extra-long hairs and their coloration of vivid black and white patterns is a warning to predators. Skunks are a major carrier of rabies over much of the USA. Rabies outbreaks, often every three to four years, cull the numbers when populations are high.

Food Mice, eggs, insects, berries and carrion form the diet of this omnivorous animal.

Young Female striped skunks produce five to six young in early May in a den lined with vegetation. Young skunks are able to spray from their anal glands at less than a month old.

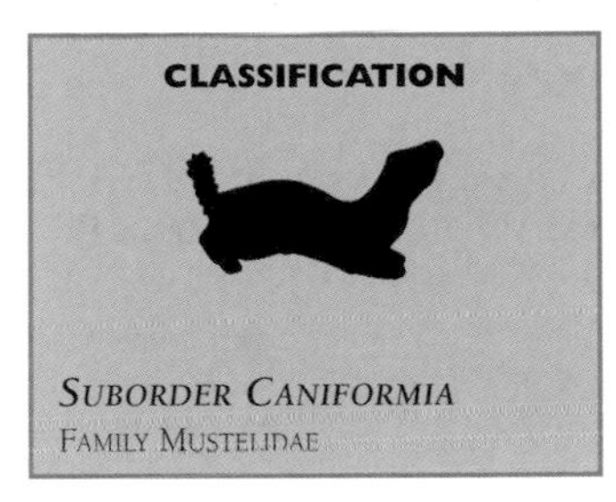

Habitat The home range stretches from southern Canada to northern Mexico, in semi-open country, woods and grassland.

MEPHITIS MEPHITIS

Male skunks take no part in the rearing of the young. In fact, they are aggressive to females with offspring and will sometimes kill the juveniles. Nor does mating occur by mutual consent. As with all the generally larger males in the family, they violently accost the females, hold them down, often with a neck bite, and copulate vigorously for long periods of time.

Spotted Hyena

Hyenas are members of the cat-like carnivore group, which are believed to have evolved from a civet-like ancestor of the Eocene. Fossil hyenas, once dominant carnivores, have been found in North America, Europe, Asia and Africa from 10 mya. Hyenas are still among the largest predators, with massive long forelegs and short hindlimbs. Few who have heard the cry of the spotted hyena in the still air of an African night will forget the presence of this fearsome animal.

Close relatives There are three true species of hyenas in the family; the fourth family member is the more lightly built aardwolf.

CROCUTA CROCUTA
Hyenas scavenge during the day and hunt live prey at night. They derive a major portion of their diet from scavenging on large ungulate kills. A group of 38 spotted hyenas have been observed dismembering a full-grown zebra in 15 minutes.

Although they look like dogs, these land-dwelling carnivores are related to viverrids and cats.

Characteristics The common names of hyenas are descriptive: the spotted, striped and brown are not easily confused. They range from 40 to 55 inches (105–140 cm) in head and body length; and 65 to 175 pounds (30–80 kg) in weight. They all have a distinctive dorsal mane and an anal pouch used to deposit scent, which is structurally unique in the feli-forms. Also differentiating

CLASSIFICATION

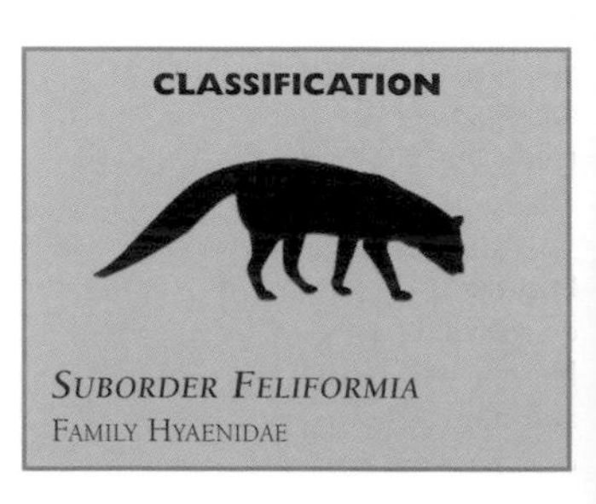

Suborder Feliformia
Family Hyaenidae

this family among the carnivores is the structure of the inner ear.

Food Carrion forms a large part of hyenas' diet but spotted hyenas are also aggressive hunters, using pack tactics to kill large animals such as young rhinoceroses or adult wildebeests. Spotted hyenas frequently bury prey in muddy pools to which they return when the are hungry.

Young Courting spotted hyenas eject strong-smelling anal-gland secretions. Males are boisterous with females, rolling them on the ground before mating. One to four offspring are born in a burrow. Their eyes are open and they already have some teeth, but they are suckled for 12–18 months.

Habitat Hyenas, in general, are creatures of open habitats. Spotted hyenas are confined to semi-desert and moist savanna in Africa, south of the Sahara.

CLEANERS OF THE SAVANNA

Hyenas must often compete with jackals and vultures before they can scavenge the remains of a big cat kill. Spotted hyenas seem to prefer bone and skin to meat even when they have killed for themselves. They have large heads with impressive high, bony crests on the top of the skull, which serve as an attachment for the massive jaw muscle that gives hyenas one of the strongest bites of any carnivore. They have robust, bone-crushing teeth, and are able to regurgitate hooves, antlers and any other matter that they cannot digest.

Indian Gray Mongoose

Mongooses are successful terrestrial carnivores that range over large distances in search of food. They are found only in Africa and southern Asia, with one species extending into southernmost Europe. Unlike their cat relatives, they do not have retractile claws and so are poorly adapted for climbing.

HERPESTES EDWARDSI

Although snakes do not predominate in the diet of any mongoose species, the mammal's speed and agility coupled with its powerful bite can overcome a cobra. In *The Jungle Book*, Rudyard Kipling described such a duel in his tale of Riki-tiki-tavi, the Indian gray mongoose.

Close relatives This large group is divided into 31 species in 17 genera.

Characteristics Mongooses are small, ranging from about 10 to 24 inches (25–60 cm) in head–body length. Unlike other feliforms, the two inner ear chambers are located one in front of the other. Mongooses are diurnal animals; nocturnal rodents are relatively safe from these predators. Indian grays, like most other mongooses, are solitary hunters and uniform in color.

Food Mongooses are famous as "ratters." They also eat insects, spiders, scorpions, centipedes, small frogs, lizards, snakes, other small mammals and birds. Fruit and plant shoots are the necessary vegetable content of their diet.

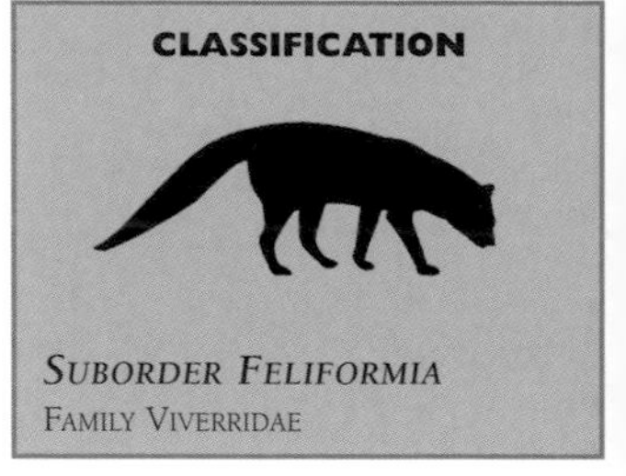

CLASSIFICATION

SUBORDER FELIFORMIA
FAMILY VIVERRIDAE

Young Litter size varies among the species, but generally two to eight young are born after six to nine months' gestation. The female raises the young alone in the solitary gray species; in more social mongooses, the young may be cared for by several females.

Habitat The Indian gray mongoose occupies east and central Arabia to Nepal, India and Sri Lanka.

MEERKAT

These animals are also known as suricates. They are easily tamed as pets and are sometimes kept to rid houses of rodents.

Close relatives Meerkats are distinguished from their close relatives, the mongooses, by having four rather than five toes on each foot.

Characteristics Meerkats have thin belly fur, which helps them to regulate their body temperature. They warm up by sitting erect and sunning themselves or lying on warm ground; they cool off by resting stomach-down in burrows or on shaded surfaces. Meerkats see and hear very well, and are constantly on the alert for birds of prey; they speedily take cover if alarmed.

Food Many foods are eaten, including insects, spiders, lizards, snakes, birds and their eggs, small mammals, fruit, roots and other plant matter.

Young Females give birth to two to five young, which are born blind in underground grass-lined chambers. Their eyes open at 12–14 days and weaning begins after three or four weeks.

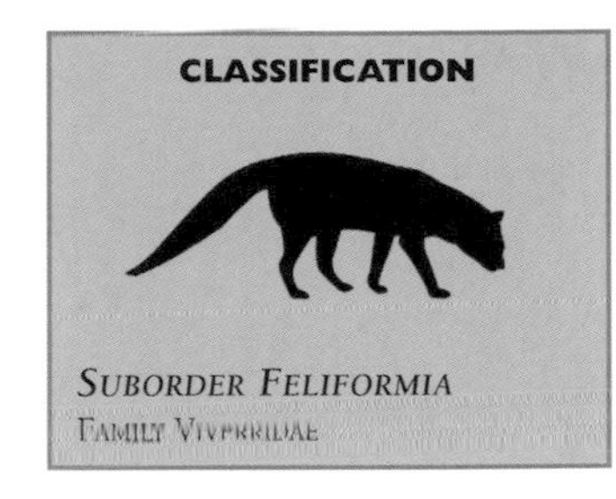

CLASSIFICATION

SUBORDER FELIFORMIA

FAMILY VIVERRIDAE

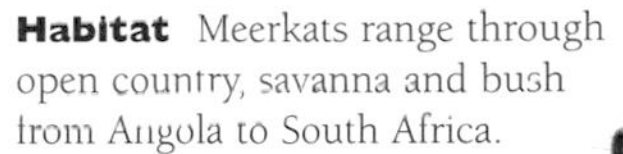

Habitat Meerkats range through open country, savanna and bush from Angola to South Africa.

SURICATA SURICATA
Meerkats are sociable, living in large communities of 30 or so animals. They dig networks of burrows or shelter in rock crevices. When food supplies dwindle, the colony moves to a different area.

AFRICAN CIVET

Civet oil from the perineal gland unique to many viverrids has been used in perfume making for centuries. The African civet is the largest true civet, belonging to the most primitive subfamily, Verrinae.

Close relatives The large family has 35 species in 20 genera and six subfamilies.

Characteristics This predatory species is a terrestrial and nocturnal ambush killer. African civets cannot retract their claws and the soles of their feet are bare between the toes and footpads.

CRYPTOPROCTA FEROX

The tree-climbing, cat-like fossa is Madagascar's largest carnivore. It is about the same size as an African civet, but is separated into a subfamily, lacking the civet gland and possessing retractile claws.

Food African civets hunt mammals up to the size of young antelopes, ground birds, reptiles, frogs, toads and insects.

Young African civets bear one to three young and may have more than one litter a year. Mothers maintain contact with their young with a chuckling contact call.

CIVETTICTIS CIVETTA

African civets begin to stir from their daytime rest in thickets and burrows at twilight. They forage at ground level, most frequently catching gerbils, spring hares and spiny mice.

Habitat African civets live in forests, savanna, plains and cultivated areas of Africa, south of the Sahara to South Africa and across the Transvaal.

CLASSIFICATION

SUBORDER FELIFORMIA

FAMILY VIVERRIDAE

CLASSIFICATION

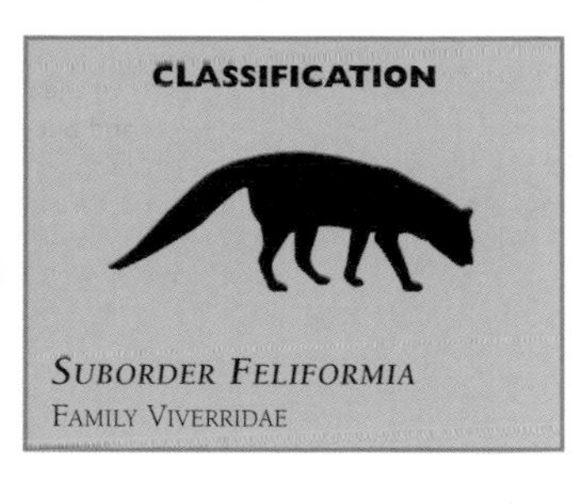

SUBORDER FELIFORMIA
FAMILY VIVERRIDAE

HEMIGALUS DERBYANUS
The banded palm civet, or banded linsang, is the best known of the subfamily Hemigalinae and is named for the broad dark vertical bands on its prettily marked body.

PALM CIVET

The palm civets demonstrate the evolution of a group of meat-eating carnivores into fruit-eaters. One species has been documented as feeding on over 30 different fruits. This somewhat elusive animal is seen only at night when it forages on the ground and through the trees. It rests during the daytime in tree hollows.

Close relatives Banded palm civets belong to the subfamily Hemigalinae of which there are five species in four genera.

Characteristics This skilled climber has strong feet, well adapted to its semi-arboreal life. It is a solitary animal with a head and body length of 21 inches (55 cm) and a tail $12\frac{1}{4}$ inches (32 cm) long. It weighs about $4\frac{1}{2}$ pounds (2 kg).

Food Earthworms and locusts are the banded palm civet's staple diet, but it also hunts for rats, lizards, frogs, crustaceans and snails.

Young There is little information about the breeding habits of this species but the litters usually contain two to three young.

Habitat Banded palm civets occur only in the rainforest of South-East Asia including the Myanmar peninsula, Malaysia, Sumatra, Borneo, Sipora and the south Pagi Islands. The masked palm civet has become a popular item in Chinese restaurants, causing alarm over its conservation status.

Tiger

The felids are the most carnivorous of the order, with sharp scissor-like teeth, horny papillae on the tongue and binocular vision. The large cats, called the pantherines; the small cats, felines; and the cheetah form subfamilies. Besides size, the main difference between them is that the small cats purr, but don't roar, and the large cats can roar, but not purr. The tiger is the largest and the only cat that may supplement its diet by attacking humans.

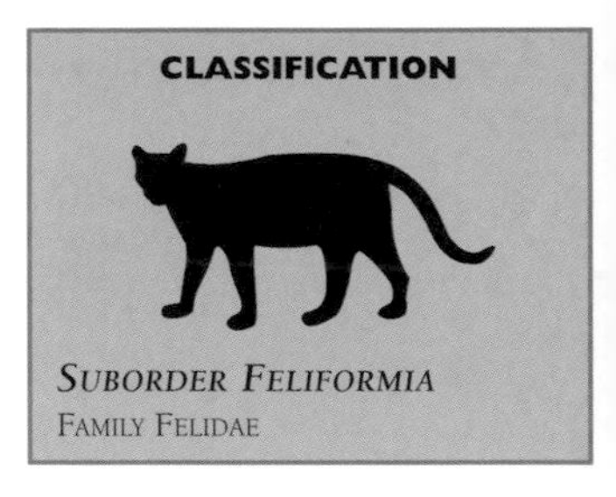

CLASSIFICATION

Suborder Feliformia

Family Felidae

PANTHERA TIGRIS

The tiger is the only felid with stripes. Its coat varies from dark orange to reddish ocher, and the belly, neck and insides of the limbs are creamy white. The Siberian race usually has the palest coat, while the tiger in Indochina is much darker. The black or brown stripes run vertically across the body.

Close relatives There are five species in the genus *Panthera*.

Characteristics The largest male Siberian tiger recorded was 790 pounds (360 kg). They are solitary and usually hunt alone, although adults may come together to share a kill. Exclusive territories are influenced by prey density. Female ranges from 6 to 386 square miles (16–1,000 sq km) are known.

Food Tigers feed on whatever large prey they can catch, including deer, pigs, gaur and buffalo.

Young Tigers mate at any time of the year, and two to three young are born after a gestation of 104–106 days. Young are nursed for six months and stay with the mother until about two years old.

Habitat Tigers live in a wide variety of habitats and climate types, but are endangered in their ranges on the Indian subcontinent, Indochina, Thailand, Sumatra and Siberia.

Lion

In contrast with the tiger, the lion is the most social of the cats, well known for its habit of living in prides. These consist of five to 15 adult females and their offspring, with one to six adult males.

Close relatives The genera *Panthera*, *Lynx* and *Neofelis* share the subfamily Pantherinae.

Characteristics Lions, which range in weight from 260 to 520 pounds (120–240 kg), hunt co-operatively and regularly kill prey above 550 pounds (250 kg). The highly visible large males do little of the killing. The adult females fan out to stalk and surround the prey, and ambush with a short burst of speed.

Food Lions feed on a variety of large and medium-sized prey, like giraffe, wildebeest, zebra and warthog, and will eat carrion and smaller prey.

Young Females in a pride will often give birth in synchrony. One to four cubs, weighing about 3 pounds (1.5 kg) are born after 110 days' gestation. Cubs can suckle from any female in the pride and remain with their mother for two years.

Habitat Lions are found in open, grassy plains, savannas, arid woodlands and semi-desert, across Africa, south of the Sahara to Botswana. A small population still survives in the Gir forest of western India.

CLASSIFICATION

SUBORDER FELIFORMIA
FAMILY FELIDAE

PANTHERA LEO
Lions show the most marked sexual dimorphism of the cats: the males have impressive manes and are 25–30 percent larger than females.

THE HUNTING LION

Lions are not only the most social of the cats, but also among the largest and most powerful. Their reputation as superlative hunters, however, is not entirely justified, as they only make a kill once in every five attempts. Lions will scavenge as much food as they can. They can catch animals as small as hares and as large as bull giraffes.

Lions and their prey Lions are opportunistic predators, taking whatever prey they can catch or scavenge. If one kind of prey becomes scarce, they will switch to another kind. It is, however, the density of prey that determines the distribution of the lion rather than the other way around. Lions may not greatly affect the numbers of their prey, but exercise selection by taking solitary animals, as well as the young, old, lame, sick, wounded and unobservant, thus influencing the evolution of the prey population.

THE STORY OF THE HUNT

Stage 1 Lions often rest near a herd of grazing animals, observing them, and waiting for the sun to set.
Stage 2 Under the cover of darkness, the females of the pride fan out to surround the herd, concealing themselves using the slightest cover—shrubs or tufts of grass—as they stalk their intended victims.
Stage 3 When the lioness has identified a suitable individual prey, she moves toward it cautiously, with eyes firmly fixed on the target. She moves stealthily with speed when the prey is preoccupied, but freezes if it looks up or shows signs of nervousness.
Stage 4 The lioness is highly alert as she comes closer to the prey, approaching with head and body close to the ground. With patience that may be tested for several hours, the lioness aims to get close enough to make a final rush.
Stage 5 If possible the lioness will spring on the back of the prey, until it buckles under her weight; or she may have to make a dash for the prey. Chases are brief, covering 990 feet (300 m) at most, in less than 20 seconds. Lions must rely on stealth rather than speed to catch their prey.
Stage 6 The lioness aims to throw the prey off its feet and with the help of other members of the pride, will subdue it as soon as possible. She rakes her forepaws over the victim's back to knock it over, or may sink the claws into its back to bring the prey down. Once down, there is little chance of escape.
Stage 7 Medium-sized prey, such as zebras, are relatively easy for a group of lions to kill, while providing enough meat for the pride to share.

The lioness separates a zebra from the herd and gathers speed to catch her prey.

She chooses her time to pounce and another member of the pride helps with the kill.

The hunting party fans out to cover as much ground as possible.

The male moves in for his share.

The lion and lioness observe the prey.

The pride benefits from the hunt, with the cubs the last to feed.

Leopard

The leopard is quite similar in appearance to the jaguar, but lacks the jaguar's massive head and robust physique. The leopard's coat is covered with small black spots and rosettes, but, unlike that of the jaguar, the open rosettes do not usually have smaller spots inside.

Close relatives There are two species of leopard within the genus *Panthera*.

Characteristics Leopards are solitary cats that maintain territories of 3–24 square miles (8–63 sq km). The head and body length is 35–75 inches (91–191 cm). They are great generalists in their habitat use, able to live in almost any area with sufficient cover and food.

Food They will eat almost anything from insects and rodents to large ungulates that are several times their own weight.

PANTHERA PARDUS
Leopards show the greatest variety in coat coloration of all the cats, both within the species and in their local populations. The background fur varies from gray to rusty brown. Desert or savanna-living leopards are usually paler, while those living in tropical forests are darker. All-black leopards are common. The exotic-looking snow leopard *Panthera uncia* has long, thick, smoke-gray fur patterned with large dark rosettes and spots.

CLASSIFICATION

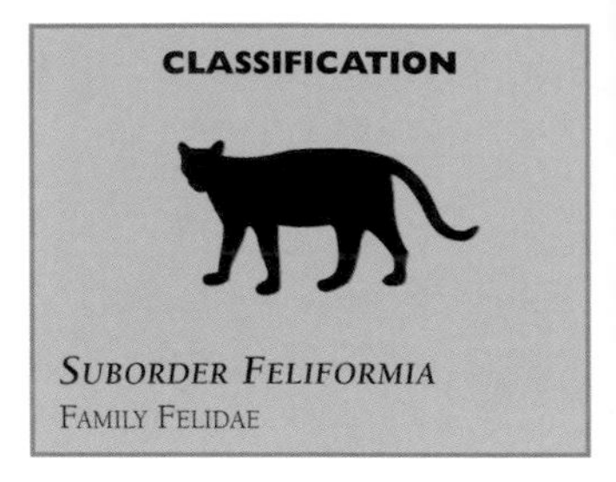

SUBORDER FELIFORMIA
FAMILY FELIDAE

Young After a gestation period of 90–105 days, two or three young are born in a secluded den. Cubs weigh 1 pound (500 g) at birth and open their eyes after 10 days.

Habitat The leopard is found in rainforest, wooded savanna, scrub, rocky mountains, desert and grazing land. It is widespread throughout Africa, southern and South-East Asia and in China and Siberia.

Jaguar

The jaguar is a powerful, deep-chested, stocky cat, with a large, rounded head and short, sturdy limbs. It looks much like a heavyset leopard, but the rosettes on a leopard's coat do not have spots inside them, as on the jaguar. The jaguar's range has been substantially reduced in the last 100 years and it is now considered endangered.

Close relatives Jaguars are one of five members of the genus *Panthera*.

Characteristics The jaguar's fur varies from pale gold to a rich rusty red. Males commonly weigh about 121 pounds (55 kg) and the females average 80 pounds (36 kg). They are solitary animals and their home range varies in response to prey density, habitat and human disturbance from 4 square miles (10 sq km) in the smallest female range, up to 60 square miles (152 sq km) for a male.

PANTHERA ONCA
The jaguar's powerful jaws and robust canine teeth enable it to kill livestock weighing four times its own weight, often with a bite to the back of the skull, rather than the more common neck or throat bite.

Food The diet includes lizards, snakes, capybara, caiman, deer, fish, turtles and cattle.

Young One to four young are born after a gestation of 93–105 days and remain with their mother for about two years.

Habitat Jaguars are often associated with well-watered areas in south-central Mexico, Central America and South America, as far as Argentina.

PUMA

Puma is an Incan word, but this cat is also known as a mountain lion, American lion or cougar, probably as a result of having the most extensive range of any terrestrial mammal in the western hemisphere. The puma's ancestry remains largely unknown, and it is one of the few plainly colored cats.

Close relatives The 25 species of *Felis* form the subfamily Felinae. However, molecular genetics may lead to alignment of the puma in the Pantherine group, or even with the now-extinct cheetah-like cats that disappeared at the end of the Pleistocene.

Characteristics Coat color and size vary greatly across the puma's broad geographical range. It is the largest cat in more than half of its range, but is shy. Pumas living within the range of the jaguar near the equator can be half the size of those living further north or south.

Food Deer are the preferred prey.

Young Two or three spotted cubs are nursed for at least three weeks, but begin to eat meat at six weeks.

Habitat Pumas live in diverse habitats from sea level to 14,765 feet (4,500 m), through western Canada and North America, and across Mexico, Central and South America.

FELIS CONCOLOR
Despite their size and strength, pumas are shy and retiring. They do not announce their presence with roars, although the calls of the female in estrus can be quite loud. Like other cats, pumas use a covert marking system of scrapes and scent marks in places that are obvious to other pumas, but not to other species.

CLASSIFICATION

SUBORDER FELIFORMIA
FAMILY FELIDAE

Cheetah

Cheetah is a word derived from the Hindu "chita" meaning "spotted one." Cheetahs have several morphological features not found in other felids: a slight build; long, thin legs; and a small, delicate skull. The cheetah is the last surviving species of at least four types of cheetahs that lived during the Pleistocene.

Close relatives The cheetah is not grouped in either subfamily of the Felidae, but remains in an undetermined position. Seven subspecies are recognized from subtle differences in their coats.

Characteristics Cheetahs are notable, not only for their beauty and unusual build, but also for their hunting technique, which involves a concealed approach toward prey, followed by a sudden rapid dash and a very fast chase of less than 990 feet (300 m). They are the most active of the cats during daytime, relying on sight to locate their prey. Cheetahs have a unique and highly flexible social structure. Females do not seem to be territorial, but avoid each other. Males sometimes join together in small groups and have a smaller home range than females.

Food The cheetah chases mammals weighing less than 88 pounds (40 kg), like gazelles, impala and hares.

Young Usually three to five, but occasionally eight cubs are born and begin to follow their mother at six weeks old.

Habitat It lives in open habitats in Africa and the Middle East.

ACINONYX JUBATUS

The speed of the fastest land mammal has been clocked at 60 miles per hour (95 km/hr). The cat expends tremendous amounts of energy during a chase, which will generally only last for 20 seconds. The cheetah's claws are exposed even when completely retracted, giving the cat grip during fast turns and rapid acceleration.

LYNXES

Members of this group are medium-sized cats, with a gray or tawny coat and varying degrees of black spots, dorsal stripes and barring on the legs. They all have a prominent facial ruff, a relatively short tail and their ears are tipped with tufts of black hair. The belly of these cats is usually heavily spotted, and this trait has accelerated demand for their pelts.

Close relatives There are five species within the *Lynx* genus: the Eurasian lynx *Lynx lynx*; the bobcat *L. rufus*; the North American lynx *L. canadensis*; the Spanish lynx *L. pardinus*; and the caracal *L. caracal*.

Characteristics Lynxes and bobcats can live for 15 years in the wild. They are resilient and adaptable solitary predators. They maintain exclusive home ranges and avoid being near adjacent home ranges at the same time, except during mating. Territorial boundaries are conveyed by scent marking, helping cats avoid injury caused by direct conflict.

Food All *Lynx* species are opportunistic feeders that have been recorded as eating 45 different food items in a year. They specialize, however, in the capture of prey the size of rabbits and hares, with rats, squirrels and mice of secondary importance.

LYNX LYNX
The Eurasian lynx is nearly twice as large as its North American relative, but with similar body proportions. Both species have large paws with thick fur padding that act as snowshoes. The Eurasian lynx is usually marked with large, well-defined dark spots. It has a ruff or collar of long hair around its neck and under its chin, as well as particularly pronounced ear tufts.

CLASSIFICATION

SUBORDER FELIFORMIA
FAMILY FELIDAE

Young Reproductive behavior is similar in all members of the genus. They come together to mate between December and April, then part company and the female is solely responsible for raising one to six kittens. Dens are usually in a cave, hollow log or under a rock ledge. Kittens' eyes open after nine days and they suckle for up to four months and stay with the mother until the next breeding season. Young males disperse and travel long distances in search of a territory, while females often settle near the range of their mother.

LYNX PARDINUS

The Spanish lynx is distinguished from the Eurasian by its smaller size and well-defined, heavy spotting. The range of the Spanish lynx once overlapped with that of the Eurasian lynx, but today it exists only in a few remote areas of the Iberian Peninsula.

Habitat The bobcat is the most successful in the group in adapting to varying habitats, and retains most of its original distribution through Mexico and the USA. Generally, other *Lynx* prefer forested habitats. The Eurasian lynx ranges through Europe and temperate Asia. The North American lynx lives in Canada and Alaska. The Spanish lynx exists in remote south-western Spain; and the caracal is found in dry areas of Africa, the Middle East, Asia Minor and India.

LYNX CANADENSIS

The North American lynx has long, thick fur and large, densely furred feet, which may measure 4 inches (10 cm) across. This tall cat has particularly long hindlegs, giving it the appearance of being tilted forward. It feeds almost exclusively on snowshoe hares.

Caracal

The long-legged caracal shares the defining features of its genus with the other *Lynx*. Like the North American lynx, it has long hindlegs that give it the appearance of being tilted forward.

Close relatives It is one of five *Lynx* species.

Characteristics The caracal is solitary and hunts mainly at night. About the size of a springer spaniel, it is the heaviest of the small African cats, weighing up to 44 pounds (20 kg). Lacking great speed, they rely on vision and hearing to locate prey, with slow, deliberate stalking. The tail, at 8–14 inches (20–35 cm) is about one-third as long as the head and body.

Food Caracals feed on birds, rodents, hyraxes and antelope fawns and are quite capable of taking large prey like impala and reedbuck.

LYNX CARACAL
The caracal's fur is short and unspotted. It is usually reddish brown on the back and sides, but white on the chin, throat and belly. The black-backed ears are topped with a long (2 inch/ 4.5 cm) tuft of black hair. All-black caracals have been recorded.

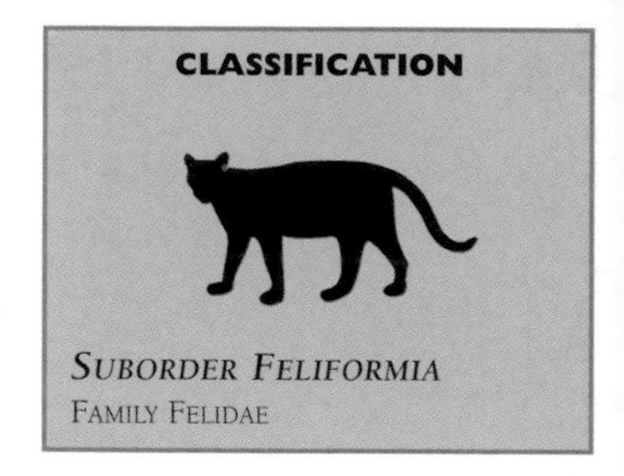

Young One to three young are born after a gestation period of 79 days. The young stay with the mother for about a year.

Habitat The caracal lives in dry woodlands, savanna, acacia scrub, arid, hilly steppe and dry mountain areas. Unlike many of its relatives, it does not live in dense forest. Its distribution is across much of Africa, parts of the Middle East, Saudi Arabia, the former USSR, Afghanistan, Pakistan and north-western and central India.

Bobcat

The bobcat is the smallest member of its genus. It generally has shorter legs and smaller feet, but otherwise is very similar in appearance to the other *Lynx*.

Close relatives The bobcat is one of the *Lynx* group.

Characteristics The bobcat is about twice the size of a domestic cat. Its fur is short, soft and dense and varies greatly in coloration from light gray to reddish brown. It has the short tail typical of the genus. It is a solitary species, though male ranges may overlap with other males.

Food Cottontail rabbit, snowshoe hare and jackrabbit are the most common food of bobcats. They will also feed on rodents, opossums, birds, snakes and deer.

Young Two to four kittens are born blind and helpless, but mature quickly. They nurse for about two months and by nine months are independent from their mother and able to catch small prey. Females can breed in the spring after their birth.

Habitat Bobcats live in coniferous and hardwood forests, brush and even deserts. Snow accumulation seems to be the main factor limiting their northern distribution. Their range extends from southern Canada to central Mexico, but they have been eradicated from some mid-western and eastern states in the USA.

LYNX RUFUS
While bobcats and lynxes spend most of their time on the ground, they will readily climb trees to escape predators like pumas, wolves and dogs. Bobcats are also strong swimmers and have been known to cross wide rivers.

Jaguarundi

The jaguarundi is one of the few unspotted cats, having an almost uniform color of red-brown or gray fur. It is also the only cat species without so-called "eye spots" of contrasting color on the back of its ears.

Close relatives The jaguarundi is a member of the subfamily Felinae.

Characteristics The jaguarundi is a solitary cat that hunts mainly on the ground during the day. It seems able to live in more open areas than many of the other neotropical felids. The jaguarundi is about the size of a large domestic cat.

Food It eats small mammals, arthropods, birds, opossums, fruit, rabbits, armadillos and monkeys.

Young Commonly two, but occasionally three kittens are born after a gestation of 72–75 days. The young are born spotted, but the markings soon fade. Kittens begin to eat solid food at about six weeks old.

Habitat This cat can live in a variety of habitats from arid thorn forests to dense second-growth forests and swampy grasslands. Its range extends from southern Texas through Mexico, Central America and into South America east of the Andes and to northern Argentina.

CLASSIFICATION

SUBORDER FELIFORMIA
FAMILY FELIDAE

FELIS YAGOUAROUNDI
A very unusual-looking felid, the jaguarundi's long-bodied, low-slung build is reminiscent of a marten; it has been compared with a weasel or an otter. This cat has a long, slender body (20–30 inches/50–77 cm), short legs and a very long tail (11–20 inches/28–50 cm). Its head is short and flattened and it has short, round ears.

Ocelot

Although recent fossil evidence for the ocelot extends through much of the southern USA, populations have declined due to hunting and habitat destruction. Ocelots and margays have very similar coat patterns and are therefore often difficult to tell apart. Ocelots are generally bigger, with shorter tails and slightly smaller eyes.

Close relatives There are 25 members of the genus *Felis* in the subfamily Felinae.

Characteristics Ocelots are medium-sized cats with short, close fur marked with both solid and open dark spots that sometimes run in lines along the body. The ears are rounded with a prominent white spot on the back. They are solitary, although males have territories that may overlap several females. They hunt in dense cover, mainly at night.

Food Ocelots feed primarily on rodents and mammals weighing less than 2 pounds (1 kg).

Young One or sometimes two kittens are born and begin to follow their mother at about two months old, dispersing from their natal range at two years old.

Habitat Ocelots are found in a broad range of tropical and subtropical habitats from southern Texas, through parts of Mexico and Central America into South America as far as Argentina.

FELIS PARDALIS

Scientists generally agree that the archetypal felid was a forest-dwelling cat. It was certainly an agile climber and leaper, which probably lived by catching rodents and other small prey in dense cover. Today, the ocelot makes its living in a similar way.

CLASSIFICATION

Suborder Feliformia

Family Felidae

Marbled Cat

This small cat is all but unknown to us. In coat color and markings it looks like a miniature clouded leopard, but with shaggier fur and a long tail. Only about six specimens are held in captivity and they are rarely seen in the wild.

Close relatives
The small cats of the genus *Felis* are relatives.

Characteristics The solitary marbled cat lives only in forested areas, where it is believed to be highly arboreal. Although its thick fur and long tail make it look larger, it is the size of an average domestic cat, weighing only 4–11 pounds (2–5 kg). Head and body length varies from 18 to 24 inches (45–60 cm) and tail length from 14 to 22 inches (35–57 cm).

Food Birds are thought to form a major part of this cat's diet. They are also thought to take squirrels, rats and frogs.

Young Four of the six captive marbled cats are not in breeding situations and we know nothing of their reproduction.

Habitat The marbled cat is distributed through Indochina, Malaysia, Sumatra and Borneo. It is endangered in northern India, Nepal, Sikkim and Assam.

FELIS MARMORATA
The background color of the marbled cat's coat can be dark brown, gray, yellowish gray or red-brown. The fur is marked with large, dark blotches, stripes and spots. In proportion to its body, it has one of the longest tails of the felids.

CLASSIFICATION

SUBORDER FELIFORMIA
FAMILY FELIDAE

SERVAL

Elegant with their long legs and large ears, servals are almost fox-like. They are designed to attain maximum height and a small slim face sits atop a very long neck.

Close relatives Servals are related to the other medium-sized felids.

Characteristics The serval uses a highly specialized "sound hunting" technique. Walking slowly through high grass, it uses its dish-like ears to focus on the rustlings of rodents. It may stop and sit for 10 minutes with eyes closed, just listening. The technique is so sensitive that a strong wind can interfere with the cat's ability to pinpoint prey, and unless extremely hungry, it will rarely bother hunting in windy conditions.

Food Servals are specialized rodent catchers, but they will also eat frogs, lizards, mole rats, small birds and insects.

Young One to three serval young grow extremely rapidly, reaching the size of their mother within about seven months. The den is usually in dense vegetation or an abandoned burrow. The young begin to eat solid food when they are a month old and acquire their permanent canine teeth at six months of age. Females with young spend twice as much time hunting as lone females.

FELIS SERVAL

Stalking through grasslands on stilt-like legs, servals are strangely [illegible] like in shape. While long legs usually mean speed, the serval is not a particularly fast runner, because it is not the legs that are long, but rather its feet. Its height provides a vantage point to "hear into" long grass, and the front feet deliver formidable blows to kill or stun prey.

CLASSIFICATION

SUBORDER FELIFORMIA
FAMILY FELIDAE

Habitat Servals prefer watered grasslands and are widely distributed in Africa, south of the Sahara.

Leopard Cat

The darkly spotted coat of the leopard cat is generally pale brown. It is able to thrive in a variety of habitats and is not intolerant of human activity.

Close relatives Relatives are the small cats in the subfamily Felinae.

Characteristics The leopard cat is the size of a tall domestic cat, weighing between 6 and 15 pounds (3–7 kg). It is solitary and mainly nocturnal, hunting both on the ground and in trees. It can also swim very well.

CLASSIFICATION

Suborder Feliformia
Family Felidae

Food It reportedly feeds on hares, rodents, reptiles, birds and fish.

FELIS BENGALENSIS
The leopard cat has a small head and narrow muzzle. It has long, rounded ears, which have a white spot on the back. There are usually four longitudinal black bands running from the forehead to behind the neck. The background color of the coat can range from bright reddish to gray and the underparts are white. The coat is marked with dark spots which may form bands or blotches.

Young Usually two to three young are born in a den, which may be a hollow tree, small cave, or hole beneath the roots of a fallen tree. Kittens open their eyes at 10 days old and reach sexual maturity at 18 months.

Habitat The leopard cat is found in South-East Asia, China and parts of the Indian subcontinent. It is equally at home in dense tropical rainforests in Sumatra and the pine forests of Manchuria.

Fishing Cat

The fishing cat readily dives, head first, into water in pursuit of fish. It seizes aquatic prey with its paws or mouth. It is strongly tied to areas of suitable wetland habitat, so while it has a broad geographical range in Asia, its distribution is quite limited.

Close relatives The fishing cat is a member of the subfamily Felinae.

Characteristics This is a powerful and robust-looking felid. A large male fishing cat may weigh 24–26 pounds (11–12 kg). It has short, coarse fur with a gray or olive-brown background with small black spots. The body is deep-chested with relatively short legs. The fishing cat's tail is unusually thick and muscular near the base and is one-third of the animal's head and body length.

Food The diet includes birds, small mammals, snakes, snails and fish.

Young Two to three young are born after a 63-day gestation period. The young suckle until six months old and reach adult size at about eight months old.

Habitat It is associated with areas of thick cover near water, in marshes, mangroves, rivers and streams in a discontinuous distribution across Asia.

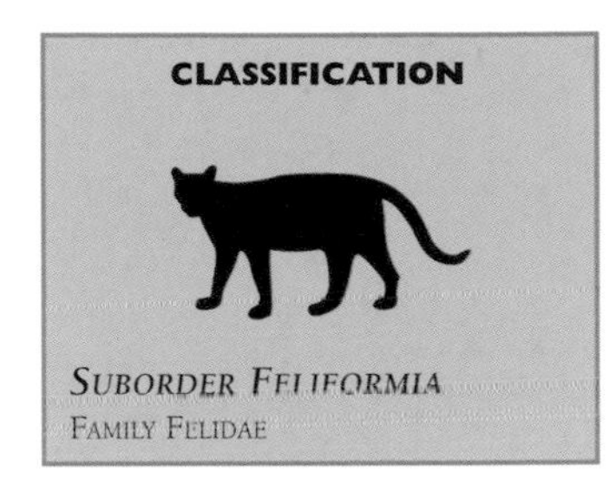

FELIS VIVERRINA
The fishing cat's powerful build and strong swimming ability enable it to take a wide range of prey. Its front toes are partially webbed and the claws protrude slightly, even when fully retracted. It is said to be able to kill calves and dogs. It has been observed crouching on rocks and sandbanks and scooping out fish with a paw.

MARGAY

The highly arboreal margay is found in forested habitats and has even been recorded living in coffee or cocoa plantations. Its color and markings are similar to the ocelot and much smaller oncilla.

Close relatives Relatives are the medium-sized felids.

Characteristics These solitary cats live at low densities, and hunt at night, mostly in trees, but sometimes on the ground. The tail is as much as 70 percent of the total length of the animal—up to 20 inches (50 cm) long on the largest individuals that are 31 inches (80 cm) in head and body length.

Food Margay eat rodents, birds, reptiles and insects.

FELIS WIEDII
The margay is anatomically adapted for arboreal hunting—with a short, rounded head, large eyes and a long tail—and capable of spectacular acrobatic feats. It is able to descend trees head first by wrapping its hindlegs around a tree trunk in a manner similar to tree squirrels.

CLASSIFICATION

SUBORDER FELIFORMIA
FAMILY FELIDAE

Young In captivity, margay give birth to a single young after a gestation period of about 81 days. The young begin to eat solid food at two months and are adult size between eight and 10 months old.

Habitat Margay can live in a variety of forest types, but are mostly associated with humid, tropical, evergreen and montane forests. It is thinly distributed from Mexico, through Central America and South America east of the Andes to Argentina.

Wild Cat

The domestic cat derived from either the European or African wild cat; African wild cats are thought to have been tamed in Egypt about 4,000 years ago. It is difficult to tell modern wild and domestic cats apart on the basis of skull morphology, although differences between skulls from last century are more apparent.

Close relatives The domestic and two wild cats are subspecies.

Characteristics Wild cats are about a third larger and more sturdily built than domestic cats. They look like a larger version of the tabby cat. Variations in body size, coat color, fur length and spotting and striping characterize wild cats as being from different parts of their range.

Food The wild cat is primarily a rodent catcher, but is a highly adaptable predator.

Felis silvestris

While it is difficult to distinguish wild and domestic cats anatomically, domestication has probably resulted in behavioral and ecological differences between them. Wild cats are solitary, territorial and seldom seen; whereas feral domestic cats form colonies and live in a diverse range of habitats in various degrees of association with humans.

Young Wild cats bear only one litter of two to three kittens a year in a secluded den. Young mature quickly and begin to hunt with their mother by 12 weeks of age

Habitat In northern Europe this cat inhabits deciduous and coniferous forests; in Africa and Asia it is found in almost every habitat from scrubby brush, to open rocky ground and even croplands.

CLASSIFICATION

SUBORDER FELIFORMIA
FAMILY FELIDAE

Sealions

The marine carnivores share a common bear-like ancestor with the terrestrial carnivores. The sealions and fur seals comprise the family of "eared seals." The family of "true" seals lack external ears. The bulky bodies and long tusks of walruses make them unmistakable as the third family of pinnipeds. Compared with fur seals, sealions have blunt noses and thinly furred coats; their skin is light-colored, rather than black. Distinguishing characteristics include a small external ear (pinna); the ability to move on all four limbs on land; and a dentition that resembles that of the terrestrial carnivores.

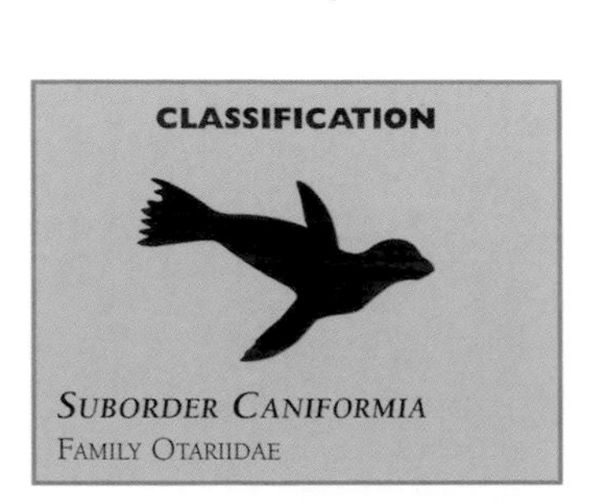

CLASSIFICATION

Suborder Caniformia

Family Otariidae

Close relatives There are five species of sealions, which form the subfamily Otariinae.

Characteristics Sealions are gregarious and like to form "rafts" floating together in the water in groups of hundreds of individuals. They drift along until suddenly, for no apparent

NEOPHOCA CINEREA

The Australian sealion is relatively sedentary, coming ashore on sandy beaches to rest, but using rocky areas for breeding. Sealions (as well as fur seals and walruses) are able to bend their flippers forward at the ankle, so their movement on land is much more four-footed than in "true" seals, which are not able to bend their hind flippers and so move on land with a humping motion.

reason, they all dive together. On adult males, the heavy, muscular neck has a mane; females are always smaller. The northern sealion is the largest species, with males up to 10 feet (3 m) in length. The coarse fur is darkish brown in males, lighter to gray in females. The Australian sealion has whitish hair on the top of its head. The Californian sealion is the most playful and barks almost continuously; males have a distinctive, prominent forehead. The southern sealion inhabits the coasts of southern Brazil and Peru, while the New Zealand sealion is the most southerly, found on the Auckland Islands.

FEMALE SEALION

Both males and females are strongly territorial during the breeding season. A strong bond grows between the female and her pup, which may last for some time after weaning. The mother stays with her pup for about two weeks, and then goes to sea to feed, returning at intervals to suckle it.

Food Sealions hunt independently in short, shallow dives for fish, squid and crustaceans. They may take penguins and carrion, and have been known to attack seal pups.

THE BREEDING ROOKERY

Sealions, as with all pinnipeds, return to the land or ice in dense associations to breed. At the beginning of the breeding season the dominant bulls arrive, select and defend their chosen territory; and try to stop cows from leaving. The pups are born shortly after the females arrive; and one to two weeks later, the females mate again, usually with the closest male. There is a two- to five-month delay before the blastocyst is implanted.

Young A single pup is born, 27–40 inches (70–100 cm) long, after an active gestation period of seven to eight months. It is fed on very rich milk and weaned at about one year.

Habitat Sealions spend most of the year at sea. They are found only in the Southern Hemisphere on and around the shores of the Pacific Ocean.

AUSTRALIAN FUR SEAL

Fur seals have dense, insulating body fur beneath a sleek covering of long guard hairs. The Australian fur seal has recovered from near annihilation in the 1820s due to indiscriminate sealing for this soft underfur and is now the most abundant of the Australian pinnipeds.

Close relatives Eight species of fur seals form the subfamily Arctocephalinae.

Characteristics Fur seals are mostly solitary outside breeding season and hunt alone. When together, females bicker and males posture and threaten. Even so, females and pups are tolerant of very dense aggregations at pupping sites. Powerful males maintain a sharply defined territory, but harems are not formed. Adult males greatly exceed female size and weight. Females weigh an average of 172 pounds (78 kg); males 615 pounds (279 kg), Their massive necks and shoulders and large canine teeth are employed in competition for breeding females.

Food Australian fur seals prey principally on fishes and cephalopods. Large prey are broken up by vigorous shaking and the indigestible remains are later regurgitated.

ARCTOCEPHALUS PUSILLUS
Australian fur seals use their four flippers to move in a slow, rambling walk or a fast gallop on land. They even climb well over rocks. They are excellent swimmers and can dive to at least 660 feet (200 m) in search of prey.

CLASSIFICATION

SUBORDER CANIFORMIA
FAMILY OTARIIDAE

Young One pup is born to each female and a maternal bond is maintained by mutual recognition of calls and odor. Australian fur seal pups are suckled for nearly a year, but the female departs to forage after mating, a week after birth. A pattern of increasing absences continues until the pup is able to accompany her at six to eight months.

Habitat Australian fur seals occur on islands and reefs in south-eastern Australian waters.

Harbor Seal

Phocids lack external ears. Their hindlimbs are directed backward and are not useful for movement on land. These "true" seals may be marine, estuarine or freshwater. The harbor seal generally rests in inshore waters on islets, rocks, sandbars and, sometimes, ice.

Close relatives Nineteen species of earless seals evolved from a stock of primitive mustelids, with which they are more closely related than eared seals. True seals are grouped as northern or southern phocids.

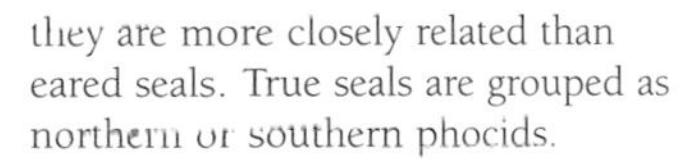

FEMALE AND PUP
Pups shed the first coat of white, woolly hair before they are born, or occassionally, soon after.

Characteristics The harbor, or common seal, is a northern phocid. Adults are relatively sedentary in habit. There are a few freshwater populations in Canada; and the harbor seal is known to wander up rivers elsewhere.

Food The harbor seal hunts bottom-dwelling and open sea fishes and some invertebrates.

PHOCA VITULINA
Harbor seals have a dog-like face and coat varying from pale to dark gray. Males have an average lifespan of 26 years, while the average for females is 32 years.

Young Births are on land between January and September. Harbor seal pups shed their white coat just before being born and are able to swim immediately.

Habitat Harbor seals are found on the coasts on either side of the North Pacific and Atlantic oceans.

Harp Seal

The harp seal is the most numerous northern pinniped, despite periods of heavy exploitation. The hunting of these animals is now controlled by the governments of Norway, Russia, Denmark and Canada.

Close relatives There are 10 species of northern phocids.

Characteristics The harp seal has a distinctive dark U-shaped harp on its back. Harp seal migrations follow the freezing and melting of the pack ice. They spend the summer in large and fast-moving groups, sometimes resting on ice, but generally haul-out only to breed and molt.

PHOCA GROENLANDICUS

Harp seals breed on ice. The mother's milk is nearly 50 percent fat and pups grow 5 pounds (2.5 kg) per day. They must lay down an insulating layer of blubber as soon as possible for survival.

Food The diet consists of fishes such as cod, krill and plankton.

Young Births are from mid-February to March. Lactation lasts for 10–14 days, after which females join underwater aggregations of vocalizing males ready for mating.

THE "WHITE-COATS"

The commercial exploitation of the appealing white-coated harp seal pups has incurred universal wrath. The coat is shed after two to three weeks for the light gray coat with dark brown spots, much like that of the adult.

Habitat Harp seals live in the sea around Labrador and Greenland, extending as far as the Arctic coast of Russia.

Hooded Seal

This species' common name arises from the curious enlargement of the nasal cavity in the male. This forms an inflatable cushion on top of the head, which increases in size with age to form a "hood."

Close relatives This northern phocid is unique in its genus.

Characteristics In both sexes, the coat is gray with large black blotches and spots, and black heads.

The hooded seal rests on ice; it rarely lands outside the breeding season. The average age of the male is 34 years, and the female, 35 years.

Food This seal feeds on squid, and fishes like halibut and redfish.

Young Loose aggregations of females give birth from mid-March to April on old, heavy ice floes. Hooded seal pups, known as blue-backs, have steely blue first coats with white fur underneath, for which they are hunted. Pups are 40–52 inches (105–130 cm) long. Lactation is 10–14 days, and the female and pup are attended by one male. In Newfoundland, the female's milk is so rich that suckling lasts for only four days.

Habitat The hooded seal inhabits the North Atlantic Ocean around Greenland and Labrador on the edge of the ice.

CLASSIFICATION

SUBORDER CANIFORMIA

FAMILY PHOCIDAE

CYSTOPHORA CRISTATA

The nasal cavity of a male hooded seal is a pouch of skin on the top of the head able to be inflated when both nostrils are closed. By closing one nostril and exhaling, the male is also able to evert the internasal membrane through the other nostril, forming a red "balloon." The function of this bizarre display is not understood.

Elephant Seals

Southern phocids are larger than northern seals, and many skull and skeletal characteristics separate the two groups. Confusingly, perhaps, the southern genera are not restricted to the Southern Hemisphere.

Close relatives The two species of elephant seals are related to the Antarctic and monk seals.

HAUL-OUT BEHAVIOR

Elephant seals alternate lengthy periods at sea with regular haul-outs on land. Individuals appear to take to the water again randomly, rather than in a mass migration. They come ashore to breed in spring, as well as in summer and fall to molt. After the breeding season, they disperse widely, largely offshore.

Characteristics The common name of these seals derives from the pendulous proboscis of the male that overhangs the mouth, so that the nostrils point downward. It is less developed in the southern species than in the northern elephant seal, and the skull of the southern is more massive. Reaching weights of 5,300 pounds (2.4 tonnes) and 16 feet (5 m) in length, male elephant seals are the largest carnivores. The southern elephant seal has the amazing capacity to remain below the surface for nearly two hours, while reaching a depth of up to 4,000 feet (1,200 m), although the average dive is

MALE AND FEMALE DIFFERENCES

The characteristic proboscis is present only in males. It becomes most obvious during mating season and when inflated, acts as a resonator for territorial roaring. The females are much smaller than the bulky males, with a maximum weight of around 770 pounds (350 kg). Females may live for 10–15 years, and males up to 25 years.

LORD OF THE HAREM
A rookery of elephant seals comprises a male with his harem and their offspring. The male spends much time and energy defending his territory, challenging other males with a bubbling roar. Older males, especially the southern species, show extensive scarring on the neck and proboscis.

20–30 minutes. Little more than two or three minutes are spent at the surface, and the animals probably sleep underwater.

Food Elephant seals prey on fishes and cephalopods.

Young A single pup is born to each female and increases in weight by 14 pounds (6 kg) daily. It is suckled for three to four weeks, and then fasts for seven weeks while it hones its aquatic skills, before going to sea at about 10 weeks.

Habitat The largest breeding populations of the southern species are on South Georgia, Kerguelen and Macquarie islands. The northern species occurs on islands off the Californian coast.

Antarctic Seals

Besides the southern elephant seal, there are four other Antarctic phocids: the Weddell, Ross, crabeater and leopard seals.

Close relatives There are nine species of southern phocids in six genera. Each Antarctic seal is unique in its genus.

Characteristics Antarctic seals do not have the territory system of the other southern phocids, elephant seals. At breeding time, animals congregate in pairs or larger groups. Blood makes up 12 percent of the body weight of seals. This carries a lot of oxygen, enabling them to stay under water for long periods. All Antarctic seals breed on the ice.

CLASSIFICATION

SUBORDER CANIFORMIA
FAMILY PHOCIDAE

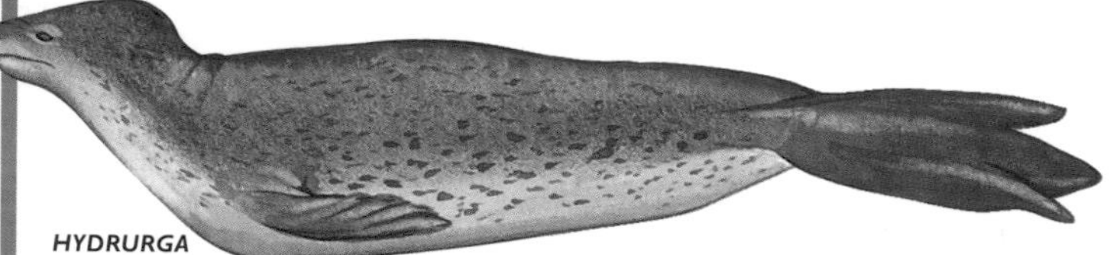

HYDRURGA LEPTONYX
Of all the pinnipeds, only the leopard seal feeds extensively on warm-blooded animals, and is the major predator of the Adelie penguin. The penguin may be seized from under water and is often tossed about. The leopard seal is known to attack young crabeater, Weddell and elephant seals.

LEPTONYCHOTES WEDDELLII
The Weddell seal swims, feeds, mates and may even sleep below fast ice, depending on holes for breathing and hauling-out. It is an excellent swimmer and the most accomplished diver of all the seals, able to descend 2,000 feet (600 m) for an hour.

Food These seals have individual preferences from the Antarctic larder: penguins, krill and other small crustaceans, fishes, cephalopods, carrion and bottom-dwelling invertebrates.

Young Phocid pups have a high daily increase in weight, suckling the fat-rich milk. Lactation lasts from one to two weeks, the timing of which is related to the stability and protectiveness of the nursery. Mating takes place toward the end of the lactation period, with a delay of two to five months before the blastocyst is implanted. Females give birth to a single pup covered in soft, thick fur that provides an insulating layer until the blubber is developed.

OMMATOPHOCA ROSSII
One of the least known of the Antarctic seals, the Ross seal is a rapid and agile swimmer, with forelimbs that are longer and more mobile than those of a typical phocid. It also has large eyes, so is well adapted to capturing fast-moving prey, like squid, under the dimly lit ice.

LOBODON CARCINOPHAGUS
The gregarious crabeater seal is the most abundant Antarctic pinniped, with a population estimated at around 50 million. It does not eat crabs, but feeds almost entirely on krill and, like the baleen whales, is a filter feeder.

Habitat The leopard seal lives in the outer fringes of the pack ice, and visits the Australian and New Zealand shores. The Weddell is the most southerly of the Antarctic seals, found on fast ice close to land. The open seas around the Pole are home to the crabeater, which may be found on pack ice. Ross seals, too, are found on heavy pack ice, with the greatest numbers in the King Haakon VII sea.

WALRUS

As the original Pacific walruses of the Oligocene died out, a group invaded the North Atlantic 7 to 8 million years ago, which dispersed and developed into the modern animals.

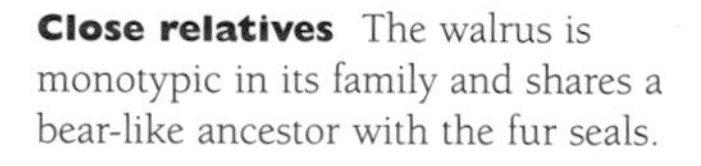

Close relatives The walrus is monotypic in its family and shares a bear-like ancestor with the fur seals.

Characteristics The bulky, cinnamon brown, wrinkled and sparsely haired body of the walrus can weigh 2,600 pounds (1.2 tonnes) and be 10 feet (3 m) long. The $1–1\frac{1}{2}$ inch (2.5–4 cm) thick skin with creases and folds at every bend of the body is a protective armor against injury from tusks. The pharynx walls in the adult male are very elastic and are expanded as a pair of pouches between the muscles of the neck. These pouches are inflated and used as buoys, and as a resonance chamber to enhance the bell-like call produced during the mating season.

ODOBENUS ROSMARUS

Like otariids, walruses are marine and able to bend their flippers forward at the ankle, so their movement on land is much more quadrupedal than in the seals. Like seals, however, the walrus lacks external ear pinna and the extensions to the hind digits are small.

Food The sensitive snout and whiskers are used to seek out mussels, clams and cockles by disturbing sediment on the sea floor. Deeper buried shells are excavated by squirting water at high pressure out of the mouth into burrows. Walruses suck the soft parts out of the mollusk shells, and also eat invertebrates and occasionally fishes.

CLASSIFICATION

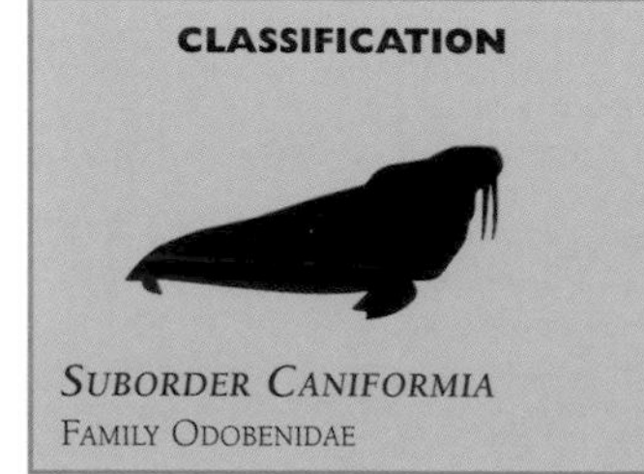

SUBORDER CANIFORMIA

FAMILY ODOBENIDAE

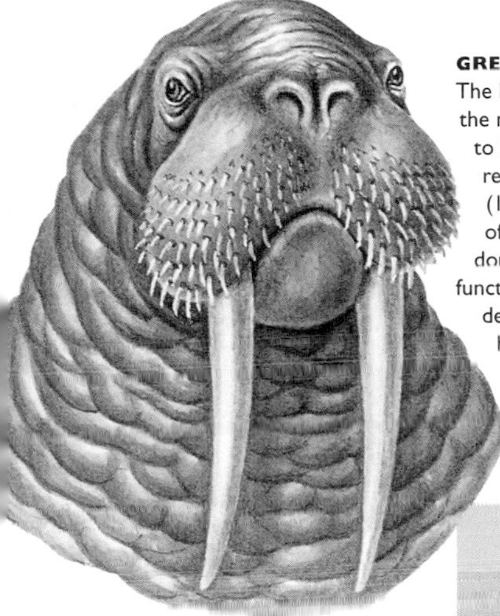

GREAT CANINES

The long tusks, present in both sexes, are the modified upper canines, which grow to 26 inches (65 cm) long, though record tusks of 40 inches (105 cm) are known. The primary role of long tusks is in establishing dominance in males. They also serve functions as diverse as ice-choppers, defensive weapons and a fifth limb to haul up onto ice. Their size and shape convey much information about the sex, age and status of an animal

Young A single calf is born every two years, after a gestation period of 15 months (including four to five months delayed implantation). Thus the walrus has a low reproduction rate. The calves are about 43 inches (108 cm) long, and suckle and remain with their mother for two years.

Habitat Walruses inhabit the seasonal pack ice in the shallower waters of the Arctic sea.

WALRUS HAUL-OUT

Walruses are highly social and vocally communicative with each other. They haul-out in vast aggregations, normally on ice, but in restricted locations will happily haul-out onto other walruses. Males segregate from females with young throughout the summer, coming together again in the winter breeding season.

Kinds of Mammals

Cetaceans and Sea Cows

The order Cetacea, which includes the biggest mammal that has ever lived, the blue whale, evolved from ungulates about 65 mya. Whales, dolphins and porpoises are now perfectly adapted to life in the sea. There are six families of toothed whales: sperm, white and beaked whales, "true" and river dolphins and porpoises; and three families of baleen whales: right and gray whales and rorquals. The forerunners of the modern order Sirenia are thought to have evolved, between 57 and 37 mya, from an ancestor shared with elephants. Four species of sea cows survive today.

THE CETACEAN WORLD

Like all mammals, cetaceans feed their young on milk; they come to the surface to breathe air through a nostril—a blowhole—on the top of their heads. They are remarkable among mammals in being fully aquatic. Their sleek, streamlined bodies and flattened tails, which propel them through the water, reflect superb adaptations to life in the sea.

Comparing whales The two suborders of Cetacea are separated on the presence of baleen plates or teeth. However, there are other physical and ecological differences between them that suggest that the baleen and toothed whales evolved independently. Great size is one: few species of toothed whales approach the length of the smallest baleen whale. Toothed whales have a single blowhole in an asymmetrical skull; baleen whales have paired nostrils in a symmetrical skull.

KILLERS AND SIEVERS

The conical teeth of toothed whales are used to seize and hold prey, but not for chewing. They vary in size and number: there are 160 in the bottlenose dolphin and 50 in the lower jaw only of the largest of the toothed whales, the sperm whale. The horny plates of the baleen whales are arranged along either side of the upper jaw, like the pages of a book, and are used to sieve plankton. The minke whale has the shortest baleen, around 3 feet (less than 1 m) in length and numbering 230–260. The 500 silky, bristly baleen of the southern right are among the largest, at around 13 feet (4 m) long.

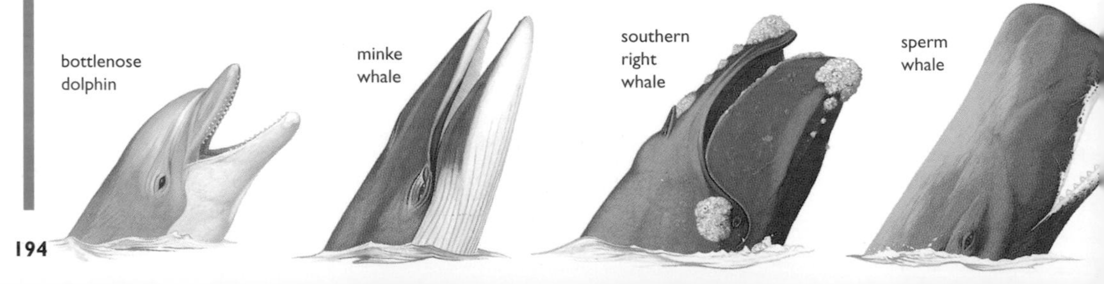

Cetacean distribution

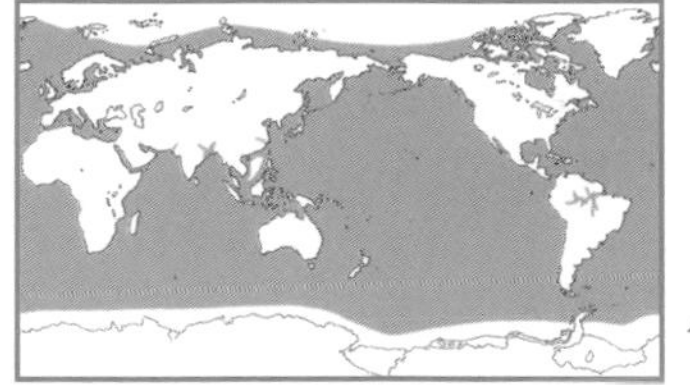

Cetaceans are found in all seas of the world and in some rivers and lakes. Many species are confined to a geographically definable area; others occur in all oceans; and many undertake extensive migrations.

Behavior The range of behavior in whales reflects their long evolutionary history and diversity of environments. Smaller toothed whales are generally faster than baleen whales, as they actively hunt their prey, using echolocation. Baleen whales tend to migrate to summer feeding areas, growing to enormous sizes and feeding on small plankton.

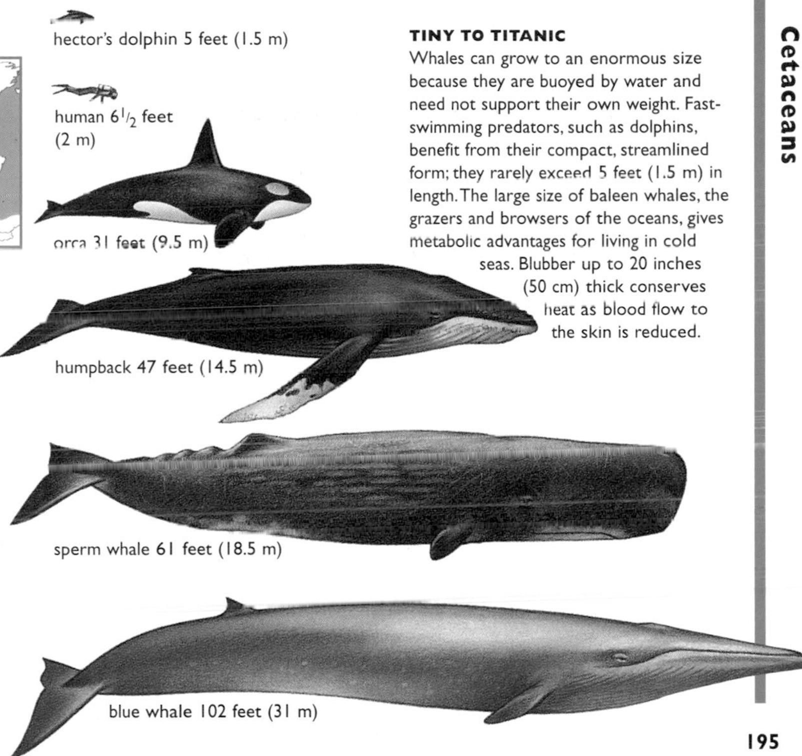

TINY TO TITANIC

Whales can grow to an enormous size because they are buoyed by water and need not support their own weight. Fast-swimming predators, such as dolphins, benefit from their compact, streamlined form; they rarely exceed 5 feet (1.5 m) in length. The large size of baleen whales, the grazers and browsers of the oceans, gives metabolic advantages for living in cold seas. Blubber up to 20 inches (50 cm) thick conserves heat as blood flow to the skin is reduced.

BALEEN WHALES

Baleen whales are the marine ecological equivalent of land-grazing ungulates. They are filter feeders, using baleen (or "whalebone") for sieving out small zooplanktonic organisms. The blue whale is the largest animal ever known to have lived. This is an impressive fact given that it, and other baleen giants, feed on some of the smallest of organisms.

Food Zooplankton are small oceanic animals, like crab larvae, krill and other crustaceans, that are found near the surface and occur in schools, enabling the baleen to capture large quantities of prey.

Zooplankton can occur in "mega-swarms" many miles across and are often heavily concentrated in sea ice.

"WHALEBONE"

The special adaptation for filter feeding in baleen whales are horny plates arranged along either side of the upper jaw, with the inner side frayed out into bristle-like fibers. The bristles may be coarse or silky depending upon the type of prey. The plates range in length from 3 to 15 feet (1–4.5 m) depending on the whale type. Baleen is a material related to keratin, like human fingernails, rather than bone; thus, the term "whalebone" may be misleading.

HUMPBACK TAIL FLUKES

The color patterns on the underside of the humpback whale's tail flukes are unique, so that individual animals are now identified and tracked as they migrate between different locations. This monitoring has provided the evidence that stocks are now increasing.

A humpback whale surfaces to blow several times prior to a feeding dive.

The flukes are raised as the whale begins its steep descent.

Size and age Baleen whales are generally large animals. The smallest, the pygmy right whale, grows to more than 20 feet (6 m), while the blue whale reaches over 100 feet (30 m) and weighs up to 180 tons (198 tonnes). Females are larger than males, perhaps to cope with the long fast required while bearing and feeding calves during their annual migrations. Some baleen whales are believed to be older than 100 years of age, so have lifespans roughly equivalent to humans.

Great travelers Most baleen whales make seasonal migrations from feeding grounds near the Poles to tropical breeding areas. Some cover a distance almost equal to half the Earth's circumference in a year. Whale migrations are not necessarily a well-defined procession of animals moving north or southward at an exact time of year. Humpback and gray whales do engage in regular migrations; other baleens take the prevailing ecological conditions into consideration.

THE FEEDING DIVE

Baleen whales feed where zooplankton are concentrated, anywhere from the surface to 300 feet (90 m) deep. They may dive vertically and surface near where they dived, or they may travel while they feed, as shown below. Feeding dives are usually short—four or five to 20 minutes—with dive length related to the depth at which food is found. A skimming motion is used that involves plowing through the water with the mouth partly open, the water streaming out from the inside through the baleen plates, with the tiny organisms being trapped on the bristles.

Bowhead

The largest, least known and once the most commercially valuable of the right whale family is the bowhead. The upper jaw is arched sharply upward, giving a mouth capacity that allows baleen plates to reach such extraordinary lengths as 15 feet (4.5 m).

Close relatives The northern and southern right whales and bowhead form the family.

Characteristics The bowhead is the stockiest baleen whale with a barrel-shaped body and enormous head that can be up to 40 percent of the 60 foot (18 m) body length. It has 325–360 baleen plates. The white chin patch unique to this species can sometimes be seen when bowheads swim upside down. It can break through sea ice to breathe.

Food The bowhead is mainly a surface feeder on very small crustaceans like copepods, steropods and krill.

CLASSIFICATION

Suborder Mysticeti

Family Balaenidae

Young Mating takes place in late summer and calves are born in spring after about 12 months' gestation. The single young is 13–15 feet (4–4.5 m) long. It is weaned after six months and calving takes place in two-year intervals.

Habitat The bowhead is the only whale confined to Arctic waters, where it follows the seasonal advance and retreat of the sea ice.

BALAENA MYSTICETUS

The bowhead whale is heavier for its body length than any other whale, estimated at around 50 tons (51 tonnes) with 28 inch (70 cm) thick blubber. It is a strong, if slow, swimmer: a harpooned bowhead once towed a fully rigged whaling ship for more than 30 hours at a steady two knots.

Southern Right Whale

The combination of the high oil yield obtained from the thick blubber layer, the great value that was placed on the fine, silky baleen, and the fact that right whales swim slowly and float when dead gave these whales their common name: they were the "right" whales to catch. They were heavily fished for 300 years and remain very rare.

EUBALAENA AUSTRALIS Breaching or leaping clear of the water is infrequent in adult southern right whales, though may be seen in the playful juveniles and young. They do not migrate as far south as the truly cold Antarctic krill-bearing waters, nor so far north for warmer calving waters.

Close relatives The southern right whale was originally considered a subspecies of its physically identical, but geographically separate northern relative.

Characteristics Distinctive callosities or skin thickenings grow on the head and lower jaw of this stocky whale. The callosity on the upper jaw is called a bonnet and is home to specialized whale lice and barnacles. It averages 50 feet (15 m) in length and the massive head is up to a quarter of the body length. As in all right whales, it lacks a dorsal fin.

Food A selective feeder that prefers crustaceans even smaller than krill.

Young Calves are born in winter after a 12-month gestation and are suckled until they reach about 28 feet (8.5 m) in length.

Habitat The southern right is widely distributed, though rare, in temperate and subpolar waters of the Southern Hemisphere.

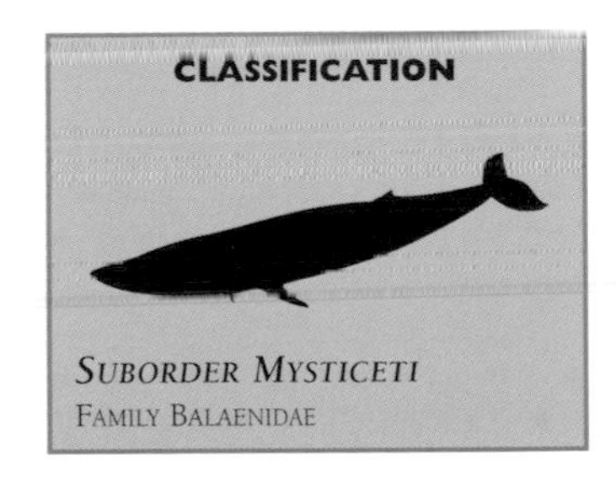

Gray Whale

Also known as the "mussel digger," this species is sufficiently unique in its appearance and feeding habits to warrant classification in a family of its own. It undertakes long-distance seasonal migrations from coastal warm-water calving areas visited in winter, to cold, temperate or subarctic feeding areas in summer. The routes are determined by the geography of the stock locations.

Close relatives This baleen whale is monotypic in its family.

Characteristics The gray is a primitive-looking species with a series of bumps along the rear third of the dorsal ridge, rather than a dorsal fin. It is mottled gray all over with patches of yellowish white or orange barnacles and whale lice. The extent of these can indicate the relative health of the individual. There is a single pair of short throat grooves, compared with none in right whales and 20–90 long pleats in rorquals. Generally one to three grays are found together, but good feeding conditions can attract hundreds.

ESCHRISCHTIUS ROBUSTUS

The gray whale uses its triangular snout to bring gammarids (related to sand-hoppers) free from the sea bottom. It then sucks up the turbid water, which is filtered through the baleen. Most grays are "right-sided feeders," stirring up the bottom with their right side, so that the baleen on that side is worn away. These whales have influenced the topography of the seabed in the Arctic, due to their plowing of bottom mud.

BEHAVIOR

The gray whale is one of the most active large whales. They have been observed spyhopping—poking their heads above the surface to eye level and then slipping back down; lobtailing—forcefully slapping the flukes against the water; breaching—surging or leaping out of the water; and surf-riding in shallow-breaking waves.

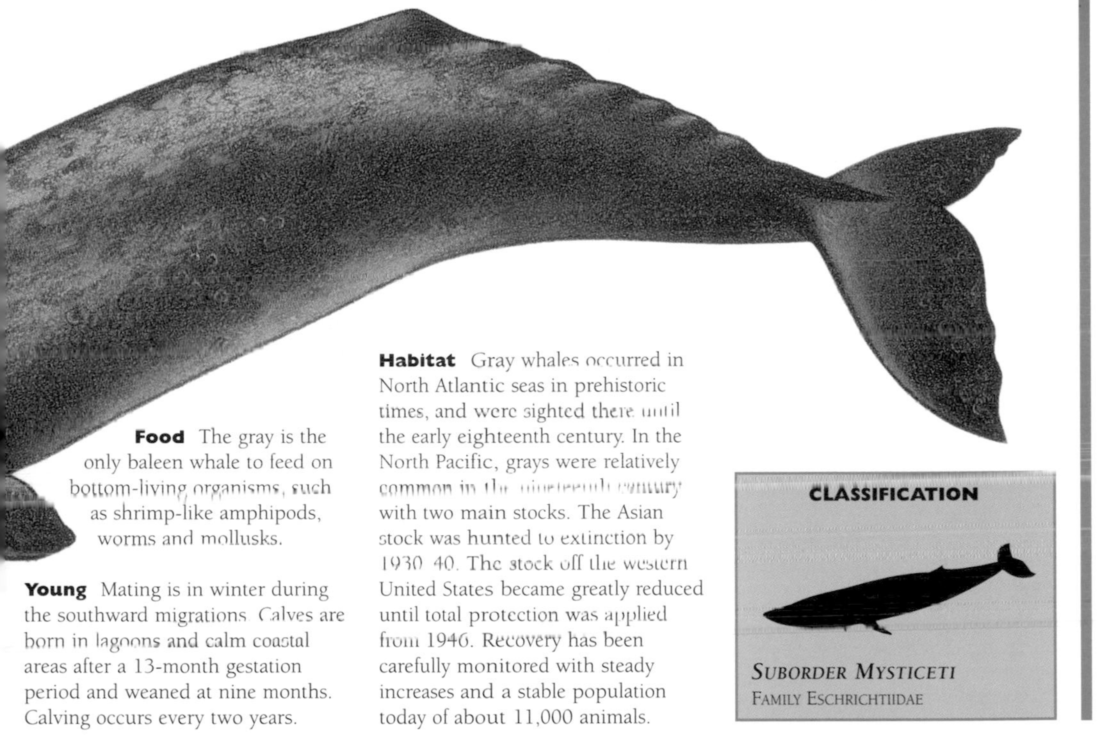

Food The gray is the only baleen whale to feed on bottom-living organisms, such as shrimp-like amphipods, worms and mollusks.

Young Mating is in winter during the southward migrations. Calves are born in lagoons and calm coastal areas after a 13-month gestation period and weaned at nine months. Calving occurs every two years.

Habitat Gray whales occurred in North Atlantic seas in prehistoric times, and were sighted there until the early eighteenth century. In the North Pacific, grays were relatively common in the nineteenth century with two main stocks. The Asian stock was hunted to extinction by 1930–40. The stock off the western United States became greatly reduced until total protection was applied from 1946. Recovery has been carefully monitored with steady increases and a stable population today of about 11,000 animals.

CLASSIFICATION

SUBORDER *MYSTICETI*
FAMILY ESCHRICHTIIDAE

Fin Whale

The six species of rorquals share the feature of a large series of longitudinal throat grooves, extending from immediately under the chin to behind the line of the flippers. Fifty to 90 pleats allow the expansion of the throat, like a concertina, to increase mouth capacity during feeding.

Close relatives Five sleek rorquals share the genus *Baelaenoptera*.

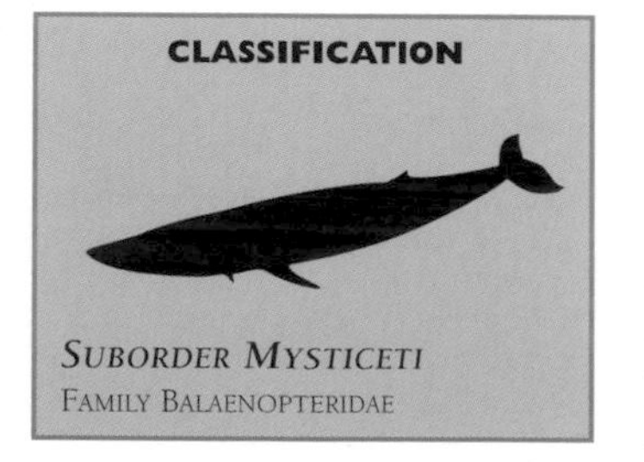

CLASSIFICATION

SUBORDER MYSTICETI
FAMILY BALAENOPTERIDAE

Characteristics A distinctive color pattern is the most notable feature of this sleek, large (45–75 ton/tonne) whale. It is dark gray to black, with the undersides white. The head has an asymmetrical dark and white pigmentation. The right baleen plates are white, while those on the left are dark. These shy animals are solitary or in groups of up to ten.

Food Small crustaceans and schooling fishes are the usual diet.

Young Calves are born at two to three year intervals in mid-winter, after an 11-month gestation. Sexual maturity is at 10 to 13 years, when males are 60 feet (18 m) and females 66 feet (20 m) long.

Habitat Fin whales are found in all oceans, from the tropics to polar waters, though they avoid shallow and coastal waters.Their flexible migration is geared to season and food supply.

BALAENOPTERA PHYSALUS

The fin whale is the second largest animal on Earth. It is known to grow to 85 feet (26 m). Newborns are 21 feet (6.5 m) long.There are 262–473 baleen plates, 28 inches (70 cm) long in each side of the upper jaw; and 56 to 100 throat grooves. It dives to depths of at least 755 feet (230 m) and is capable of speeds over 20 miles per hour (30 km/h).

MINKE WHALE

The smallest rorqual is graceful with a sharply pointed head that it pushes up through cracks in the ice floes. Its distinctive dive sequence distinguishes the minke whale in the open ocean. It is the most abundant rorqual and the only baleen whale being hunted commercially today.

Close relatives The minke is highly variable in appearance and some authorities recognize three subspecies.

Characteristics Minkes are primarily inshore, solitary whales, but they are sometimes seen at sea, where they breach spectacularly, exposing the entire body. The back is black merging to dark gray with a white undersurface. There are 50–70 throat grooves and 231–360 creamy white baleen plates.

Food Small schooling fishes, squid and crustaceans are taken.

Young Single calves are around 9 feet (2.8 m) long and suckle for less than six months.

Habitat Minkes are found in coastal temperate waters of all oceans. There are distinct populations in the Southern Hemisphere, the North Pacific and North Atlantic oceans.

CLASSIFICATION

SUBORDER *MYSTICETI*

FAMILY BALAENOPTERIDAE

BALAENOPTERA ACUTOROSTRATA

At a distance the minke whale could easily be confused with other rorquals, but its dive sequence is unique. The snout breaks the surface first at a sharp angle, which is then aligned with the surface as the blowholes appear. Like the sei whale, the blowholes and dorsal fin are visible simultaneously, but then the back and tail arch much more strongly than in the sei, in preparation for a long dive.

BLUE WHALE

The heart of the blue whale is the size of a small car. Its mouth can be 20 feet (6 m) long. Calves weigh 5,300 pounds (2.4 tonnes), about half as much as an adult African elephant. The blue whale is larger than any dinosaur that ever existed, yet its preferred food is inches long.

Close relatives The blue whale is most closely related to the four other rorquals in its genus.

MOTHER AND CALF
Blue whales are not gregarious animals, but are mostly solitary. Paired animals are often a mother and calf that will be suckled for less than a year.

CLASSIFICATION

SUBORDER MYSTICETI
FAMILY BALAENOPTERIDAE

Food Small krill are eaten in the Southern Hemisphere, supplemented by schooling fishes and crustaceans in the north.

Characteristics The largest blue whale ever captured was 110 feet (33 m) long. The blue-gray body is slender and an extremely small, sickle-shaped fin, located well down the back, marks the end of a ridge extending from the blowhole. There are more than 300 baleen plates and a highly distensible throat with over 40 grooves.

Young Mating is probably a temporary arrangement, occurring in warm waters; calves are produced every two to three years.

Habitat Blue whales are found in all oceans, in small populations.

BALAENOPTERA MUSCULUS
These huge animals could produce large amounts of oil and with improving technology became the prime target of twentieth century whalers. The peak year was 1930 when nearly 30,000 blue whales were taken.

Humpback Whale

The great, wing-like flippers of the humpback are up to a third of the whale's total length. It is very acrobatic and will back-somersault when breaching. The humpback produces the longest, most complex sound sequences of any animal.

Close relatives The humpback is sufficiently different from other rorquals to be monotypic in a genus.

Characteristics The head is massive and marked with protruberances containing hair follicles and sites for barnacles and whale lice. The huge fins are uniquely serrated on the leading edge. There are fewer throat grooves than in other rorquals (about 20), but the throat expands greatly. Humpbacks have been known to breach over 200 times.

Food Humpbacks feed on krill, eaten only in cold waters, and fishes.

Young Calves are born in winter every two or three years. Humpbacks are found almost to the pack ice during feeding, so undertake long migrations to warm breeding areas.

Habitat These relatively shallow-water whales migrate thousands of miles each year across the world's oceans.

MEGAPTERA NOVAEANGLIAE

Humpbacks employ a great variety of feeding methods. They may swim on their side in a circle with their mouth partly open and skim feed. Lunge feeding involves swimming to the surface at a marked angle, sometimes emerging almost vertically before closing the mouth and forcing out the contained volume of water. Four to six whales may herd schools cooperatively to feed in this way. In "bubblenetting," the whale circles below a concentration of food and, while slowly rising to the surface, expels air which rises as bubbles, trapping the animals within.

CLASSIFICATION

SUBORDER MYSTICETI
FAMILY BALAENOPTERIDAE

Sperm Whale

PHYSETER MACROCEPHALUS

Separated from the baleen whales by the obvious distinction of possessing teeth, the toothed whales are generally fast swimmers after fishes and squid. They evolved from the earliest whales more than 45 mya. Some highly evolved species of toothed whales have lost most of their teeth, developed horny thickenings on the palate or developed specialized teeth, like the narwhal. The sperm whale has 18–25 conical teeth in each side of the lower jaw, which fit into sockets in the upper jaw.

Deep-water giant squid form a major part of the sperm whale's diet. The longest recorded dive was 2 hours and 18 minutes. So, a dive will go to great depths in total darkness under extreme pressure.

Food The diet is mainly squid.

Young This slow-breeding species suckles young for two years, after a 14–15-month gestation.

Habitat Distribution is dependent upon season and sexual and social status. It lives in the deep water of all oceans except polar ice fields.

Characteristics In the front of the head of all sperm whales is a spermaceti organ containing a specialized, wax-like substance, which probably plays a role in adjusting buoyancy with changing pressures during long dives. Sperm whales routinely dive to 2,600 feet (800 m), but dives nearly 2 miles (3 km) deep are known.

Close relatives There are three species in the family: the "great," pygmy and dwarf sperm whales.

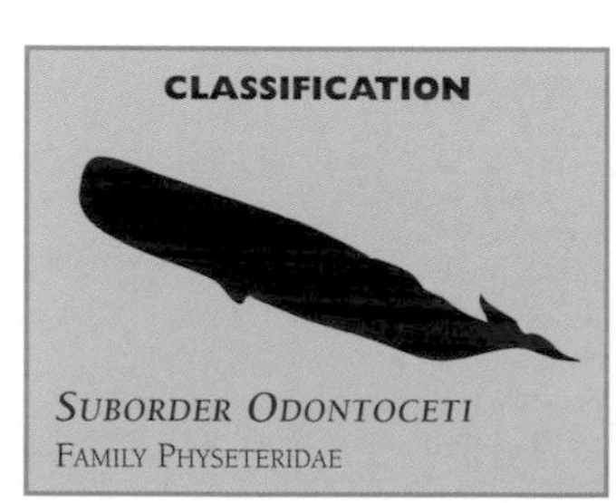

Pygmy Sperm Whale

A large, bulbous snout gives this robust whale its shark-like appearance. Like all sperm whales, the top of the head projects well beyond the tip of the narrow lower jaw.

Close relatives The dwarf sperm whale is very similar in appearance to the pygmy sperm.

Characteristics The spermaceti organ is smaller in scale, but of similar chemical structure as in the "great" sperm whale. Unlike the sperm whale, the pygmy sperm has a small fin; there is no size difference between the sexes; and they do not form schools, but are mostly solitary. Pygmy sperm whales are also deep divers. There are 10–16 long, curved teeth in each side of the lower jaw.

Food It primarily feeds on oceanic squid, but small fishes, crabs and other invertebrates are taken

Young Calves are born in late spring after 11 months' gestation. Sexual maturity is reached at 9–10 feet (2.8–3 m).

Habitat Mainly a deep-water species, the pygmy sperm whale prefers warmer waters beyond the edge of the continental shelf. Most information on the distribution of this inconspicuous whale is from strandings and it is not known whether these reflect isolated population distributions.

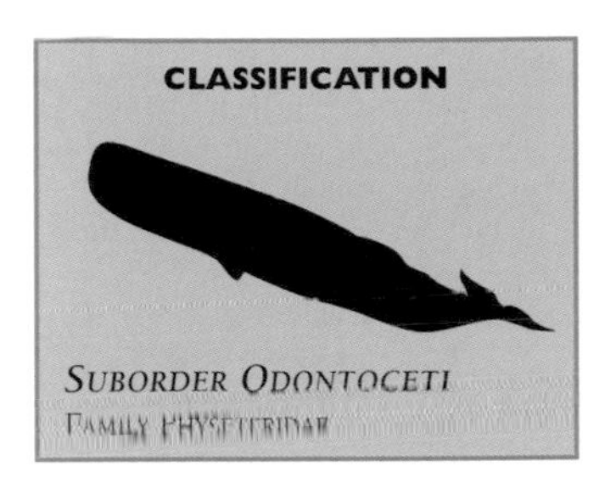

KOGIA BREVICEPS
The pygmy sperm whale rises to the surface slowly and deliberately, floating motionless with the tail hanging down in the water. Unlike most other small (9–11 foot/2.8–3.4 m) whales it simply drops quietly out of sight.

Narwhal

The white whale family includes the narwhal and the beluga. They have a blunt head, no beak, relatively long neckbones that allow head movement and a visible external neck. The famous tusk of the narwhal is apparently used in fights to establish dominance, but is the cause of its continued hunting by humans.

Close relatives White whales are most closely related to porpoises and dolphins.

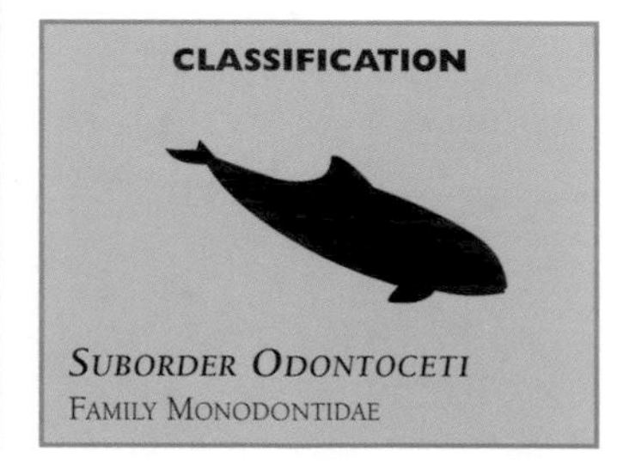

MONODON MONOCERUS

Narwhals are killed by orcas, polar bears and by becoming trapped under the ice, but their greatest threat is people. Native peoples of the Arctic hunt narwhals throughout their range—both legally and illegally.

Characteristics

Narwhals form small family pods of three to 20 individuals and larger aggregations of up to 200 move seasonally with the spread of sea ice. The modified tooth that erupts from the upper jaw in males may grow to 9 feet (2.8 m) long. On rare occasions the right tooth or both may erupt to form tusks. The bulbous forehead is used to break through sea ice and form breathing holes. The tail flukes change in shape as the animal grows, gradually becoming semicircular with age.

Food It eats squid and shrimp, and some cod and halibut fish.

Young Mating probably takes place in the Arctic spring and calves are born after 15 months' gestation. Females give birth every three years; and young males' tusks erupt at about one year of age.

Habitat Narwhals are exclusively Arctic, inhabiting deep fiords from the central Canadian Arctic to Greenland and following the seasonal advance and retreat of the ice.

Beluga

The young of this almost pure white whale are a dark brownish gray in color. They commonly swim upside-down while calling and are also known as "sea canaries."

Close relatives The beluga and narwhal comprise the family of white whales.

Characteristics The seasonal movements of belugas are highly variable: in winter they tend to form small groups, but in summer gather in bays in herds of thousands and swim hundreds of miles up major rivers. Their diverse range of sounds include loud clicks, yelps, squeaks and shrill whistles. Belugas usually swim slowly and dive for about five minutes, but can remain underwater for 15 minutes and travel nearly 2 miles (3 km) before surfacing. This ability and their sophisticated echolocation system suits them to an environment of frozen sea ice.

Food A broad range of fishes, especially salmon, crustaceans and octopus, are eaten.

Reproduction Belugas reproduce every three years or so, with a gestation period of 14 months. Calves are born in spring or summer and suckled for 20 months. Females reach sexual maturity at about five years of age; males at eight or nine.

Habitat Belugas and narwhals have complementary distributions, belugas preferring shallow water and narwhals usually the deeper water.

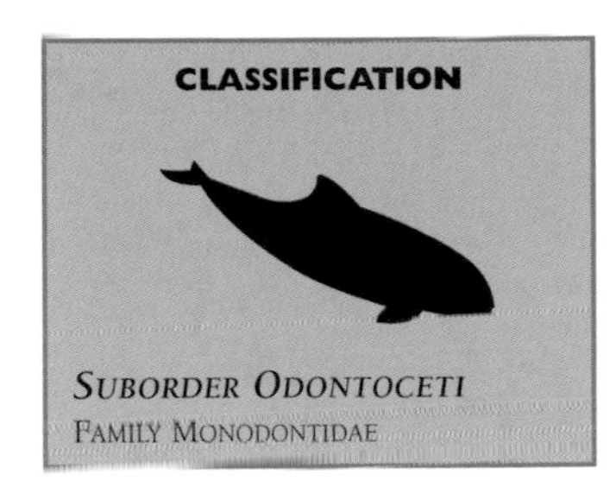

CLASSIFICATION

SUBORDER ODONTOCETI
FAMILY MONODONTIDAE

DELPHINAPTERUS LEUCAS
Belugas have unusually flexible bodies, which reflect a unique bone structure. They can rotate their flippers and heads, twist their bodies around while swimming and scull backward using the flukes. Belugas are inquisitive and spyhop out of the water, possibly to have a look around, but almost never breach.

Bottlenose Dolphin

The main difference between whales, dolphins and porpoises is size, though there is a great deal of overlap. The large baleen whales are easily identifiable as such, and no porpoise is bigger than any whale. However, among the toothed whales, some dolphins are larger than some species of whales, and some porpoises are larger than the smallest dolphins. The dolphins are the largest and most diverse family of cetaceans.

Close relatives There are 26 species of oceanic dolphins and six toothed whales in the family.

TURSIOPS TRUNCATUS
The remarkably social bottlenose dolphin displays one puzzling anomaly: the solitary dolphin. These individuals may have been ostracized, or may have chosen to live outside their society. The advantages of co-operative feeding and protection from predators are not available to them.

Characteristics The well-known bottlenose dolphin is the largest of the beaked dolphins. These powerful swimmers are mostly seen in schools of up to 30, leaping from the water, body-surfing or bow-riding. Females form the permanent basis of a group, which changes frequently as individuals come and go. The dominant, largest males form the most enduring alliances, with two or more feeding, traveling and socializing together.

Food A generalized feeder on bottom-dwelling and schooling fish, eels, squid and shrimp.

Young Females mate only every three to four years and male alliances chase females prior to mating. Gestation is one year and calves suckle for 18 months.

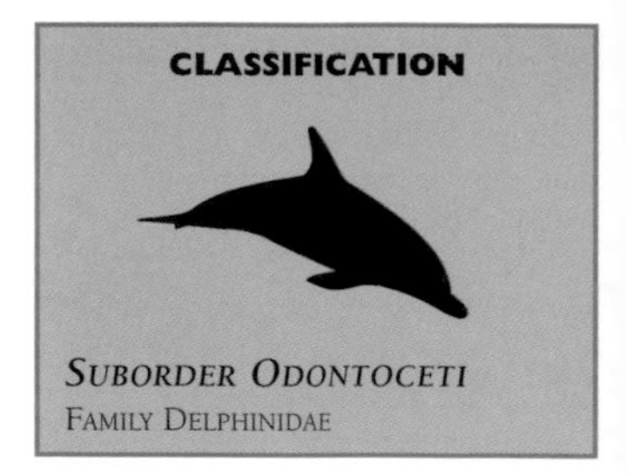

Habitat This species is widely distributed in tropical and temperate waters.

Common Dolphin

Also confusingly named the white-belly porpoise, the common dolphin is highly visible as it swims rapidly at the surface and is strikingly patterned. Celebrated in art since ancient times, it is the best known of the small whales.

Close relatives The common dolphin is monotypic in its genus.

Characteristics Common dolphins form pods of tens to many hundred. They are efficient group hunters and it is thought that their success at finding prey induces tuna to follow them. This natural association is exploited by fisheries and staggering numbers of dolphins have been killed in tuna catches. Attempts to reduce dolphin mortality have resulted in changes to nets to allow dolphins to escape more readily. Experienced dolphins have altered their own behavior to flee from approaching boats; or when trapped, to wait, apparently calmly, near the section of the net where they will be released.

Food Common dolphins feed on anchovies, herring and sardines, and on squid that are hunted at depths of up to 920 feet (280 m).

Young Sexual maturity is reached at three to four years of age, or about 6 feet (1.8 m) in length. Calves are born in spring after 11 months' gestation and suckled for six months.

Habitat It inhabits deep offshore warm waters in tropical and temperate areas.

DELPHINUS DELPHIS

Abundant and with a high rate of reproduction, common dolphins appear able to weather the pressures of deaths from pollution, commercial fishing and limited hunting. They associate with tuna and are captured "incidentally" in purse seine nets.

CLASSIFICATION

SUBORDER ODONTOCETI
FAMILY DELPHINIDAE

Orca

The appearance of the orca, or killer whale, is unmistakable: the largest dolphin with a robust and graceful body, boldly marked in glossy black and white. It is named for its predatory skill, hunting co-operatively in groups of two to 20 individuals that will harry and kill baleen whales much bigger than themselves.

ORCINUS ORCA

Orcas produce three different types of sound: echolocation clicks, whistles and pulsed calls. Analysis of pulsed calls has revealed that every pod has its own dialect. Calls remain stable over several years and different communities share very few calls.

Close relatives The orca is one of the six toothed whales in the family.

Characteristics Orcas live in relatively stable pods of up to 50 related animals (though about 30 known pods number 260 or so). Pods regularly combine with others of the same "community." Resident pods have a range of at least 310 miles (500 km). Cohesion within the pod is maintained by constant vocal communication. Orcas are among the more sexually dimorphic whales. Males are heavier and longer than females with a sharply triangular dorsal fin that may stand $6^{1}/_{2}$ feet (2 m) high.

Food The diet is extremely diverse, including seabirds, turtles, sharks, other cetaceans, seals and sealions.

Young The breeding season lasts all year. Calves are $6^{1}/_{2}$–8 feet (2–2.5 m) long at birth and are suckled for at least a year. It is thought that males may remain in the group into which they were born for life. Females mate with males from adjoining pods.

Habitat Orcas are found in tropical, temperate and polar waters, but prefer cooler, coastal areas.

CLASSIFICATION

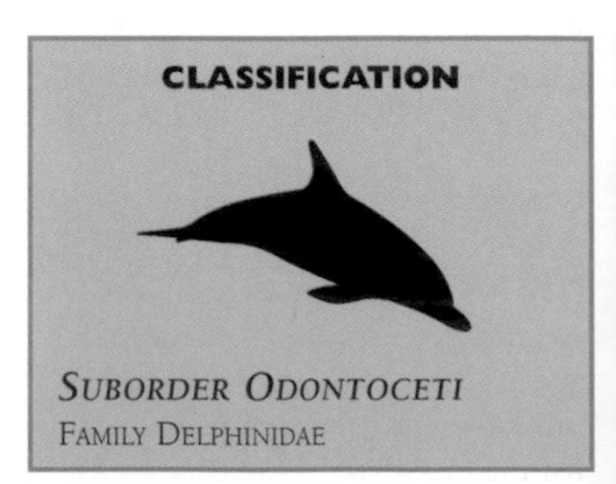

SUBORDER ODONTOCETI
FAMILY DELPHINIDAE

Spinner Dolphin

STENELLA LONGIROSTRIS

Spinning, leaping and breaching abilities, common to many cetaceans, reach their pinnacle in spinner dolphins. Combined schools of spinner and spotted dolphins work to the advantage of both species: spinners rest during the day and feed at night, while spotted dolphins rest at night and feed during daylight.

Close relatives There are several distinct forms of this species, differing in size, body shape and color. The striped and spotted dolphins are its closest relatives.

Characteristics The spinner is a particularly slender dolphin with a long, thin beak. It is an agile fast swimmer, but most information about this species has come from the millions of spinners killed in tuna-fishing operations. The eastern Pacific stock of spinner dolphins has been the most heavily exploited dolphin and has been reduced to 17 percent of its pre-1959 numbers.

In daylight hours, spinner dolphins rest close inshore in bays, swimming slowly in groups of 20 or so. They form groups that may number several hundred to search for deep-water fish at night. The transition from rest to active travelling and feeding is quite marked. As evening draws near, the aerial activity of spinners increases and is followed by a period of zig-zag swimming, apparently to ensure every member of the group is acting in a cohesive fashion.

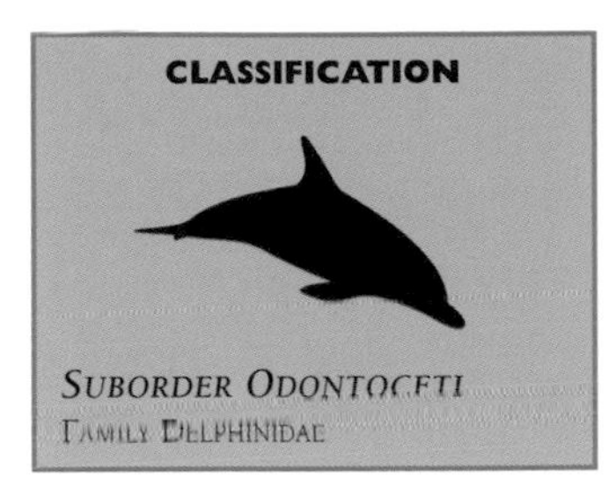

Food Prefers squid and fish found to at least 200 feet (60 m) deep.

Young A single young is born annually, estimated to be 31 inches (80 cm) long, after a gestation period thought to be around 10–11 months.

Habitat Spinners inhabit the tropical and warm temperate waters in the Atlantic, Pacific and Indian oceans. Some populations live close to the shore, while others are apparently exclusively oceanic.

Harbor Porpoise

The harbor porpoise is typical of the porpoise family, being shy and avoiding boats; rarely leaping out of the water; and swimming singly or in small groups. As a group, porpoises are smaller than most dolphins with no distinctive beak and a forehead that slopes uniformly forward to the tip of the snout.

Close relatives Burmeister's porpoise and the vaquita are its closest relatives.

PHOCOENA PHOCOENA
The harbor or common porpoise is neither common, nor often seen in harbors. It swims slowly, but if necessary, can reach speeds of up to 14 miles (22 km) per hour. Its other common name, "puffing pig," comes from the sharp, puffing sound of the blow, somewhat like a sneeze, that is rarely seen, but can be heard.

Characteristics The harbor porpoise usually only exposes its back on the water surface and dives for up to four minutes. It rarely grows larger than 6½ feet (2 m). A series of small, blunt spines on the leading edge of the dorsal fin is the main feature used to identify this species. There are 21–25 pairs of spade-like teeth in the lower jaw and 22–28 in the upper jaw.

Food The harbor porpoise catches herring, cod and other schooling fishes, bottom fishes to 300 feet (90 m), squid and shrimp.

CLASSIFICATION

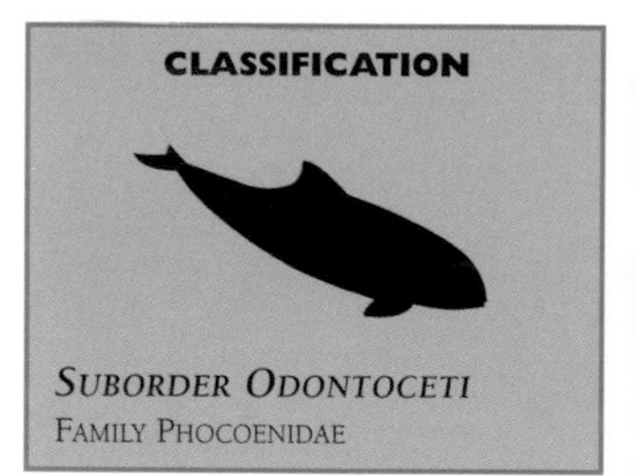

Suborder Odontoceti
Family Phocoenidae

Young Both sexes are sexually mature when about 5 feet (1.5 m) long. The young are born during the summer after a gestation period of 11 months and are 27–35 inches (68–91cm) long. Females generally calve each year.

Habitat This species inhabits bays and estuaries with murky waters caused by tidal flows, in temperate and subarctic waters of the Northern Hemisphere.

Dall's Porpoise

One of the most highly evolved porpoises is the striking Dall's porpoise. It has an increased number of vertebrae that are extremely compressed, allowing greater trunk muscle mass, and making the Dall's porpoise the fastest, strongest swimmer of all the cetaceans.

Close relatives Dall's porpoise is one of six species in the family.

Characteristics With an almost perfectly hydrodynamic shape, Dall's porpoise is best known for its astonishing bursts of speed (up to 30 miles per hour or 50 km/hr), when it produces "rooster tails" of spray behind it. The skull and air sinuses are highly evolved and its diminutive teeth are recessed among horny "gum teeth," which serve the functions of the once larger teeth.

PHOCOENOIDES DALLI

The population of Dall's porpoises is estimated at about 920,000, but it has been severely depleted during the past 25 years as a direct result of various Japanese, Taiwanese and South Korean fishing operations. Perhaps 20,000 a year are killed in salmon and squid nets and more than 10,000 a year are killed off Japan for human consumption.

Food This species eats squid and schooling fish, commonly at depths of 600 feet (180 m) or more.

Young The calf is 37–40 inches (95–105 cm) long at birth. It is generally born in summer, after a gestation period of around 11–12 months.

Habitat Dall's porpoise inhabits cold waters from Japan and California to the Bering Sea. It is mainly found in inshore waters, but is sometimes captured by tuna fishermen up to 620 miles (1,000 km) offshore.

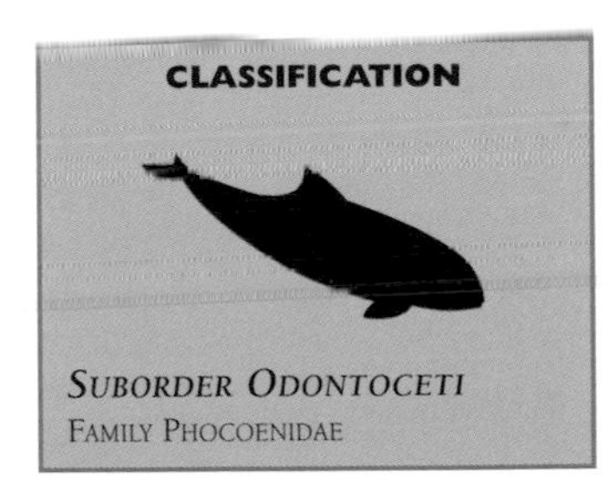

CLASSIFICATION

Suborder Odontoceti

Family Phocoenidae

Yangtze River Dolphin

River dolphins are highly specialized animals, superbly adapted to life in turbid, murky water. They are vulnerable to floods, damming, pollution, over-fishing of their prey and human predation. The Yangtze river dolphin is one of the most endangered species of toothed whales and is represented by a mere 250 to 300 individuals in the Yangtze River.

Close relatives There are five species of river dolphins in Asia and South America.

Characteristics The Yangtze river dolphin has a highly developed echolocation faculty in response to the silty waters of its habitat. The low, triangular dorsal fin is often all that is seen of this shy animal. It dives briefly. It is sometimes considered a relative of the Amazon river dolphin and shares the relatively broad, wrinkled teeth of that species.

Food Eel-like, bottom-dwelling catfishes and other freshwater fishes have been found in the stomachs of animals accidentally hooked by fishermen.

Young Males reach sexual maturity at around four years; females at six years. It is thought that calves are born between February and April, after a short gestation period.

Habitat This species is now restricted to the middle and lower reaches of the Yangtze River of China. It prefers slow-moving river areas with established sandbars.

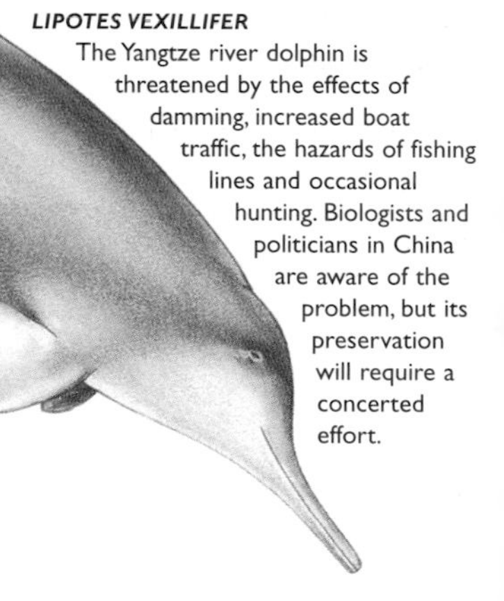

LIPOTES VEXILLIFER

The Yangtze river dolphin is threatened by the effects of damming, increased boat traffic, the hazards of fishing lines and occasional hunting. Biologists and politicians in China are aware of the problem, but its preservation will require a concerted effort.

CLASSIFICATION

Suborder Odontoceti

Family Platanistidae

Ganges River Dolphin

The Ganges "susu" is also known as the side-swimming dolphin, from its habit of swimming on its side to allow the leading edges of its large, rounded pectoral fins to comb the bottom. The eyes, which are capable only of distinguishing light from dark, are tiny and effectively non-functional—this species moves and feeds using only echolocation.

Close relatives The Indus river dolphin of Pakistan is separated geographically and is classified as a different species, but is essentially the same animal.

Characteristics Susus have extensive and complex air sinuses in the skull, which also has a large crest on each side over the eyes, giving the animal a distinctly bulbous forehead. These sightless creatures live in muddy waters using echolocation and probing with their snouts to find food on the bottom. The very long, slender beak contains 26–39 backward-curving teeth. The anterior teeth are longer than the rear and these extend outside the closed mouth like a pair of forceps.

Food The diet includes shrimp, catfish and carp.

Young Calves are born in the spring after nine months' gestation and are weaned within a year.

Habitat Widespread along most of the Ganges, Brahmaputra and Karna-phuli rivers.

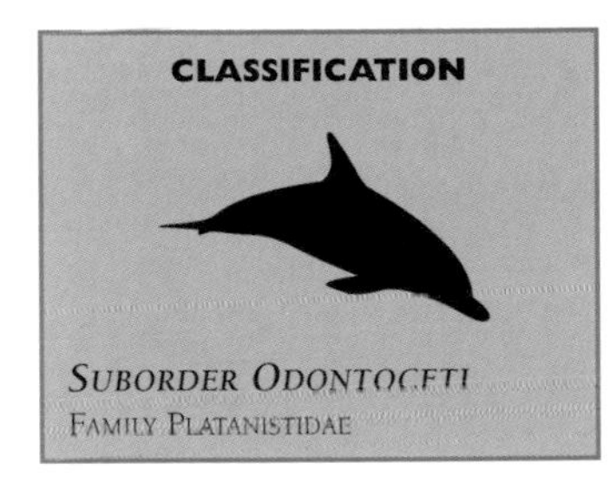

Platanista gangetica

Susus are slow breeders and are under considerable threat in their vulnerable habitat, which has been largely destroyed by environmental degradation and damming. Illegal hunting also continues.

Sea Cows

Entirely unrelated to seals and whales, sea cows (order Sirenia) are unique marine mammals in being herbivorous. They evolved during the Miocene when conditions favored the growth of freshwater plants in the rivers along the South American coast. The silica in these plants rapidly wears away teeth, so manatees evolved a system of continual replacement of their molar teeth. The dugong has developed open-rooted molars that continue to grow throughout life.

Close relatives There are four living species of sea cows: one dugong (Dugongidae) and three manatees (Trichechidae). A fifth species, Steller's sea cow (Hydrodamalidae), was exterminated in the eighteenth century.

Characteristics As aquatic herbivores, sirenians have characteristics of both marine mammals and terrestrial herbivores. The dugong's body shape and tail flukes are like those of a dolphin. The small external ear holes also resemble those of cetaceans; its hearing is acute. The dugong's

CLASSIFICATION

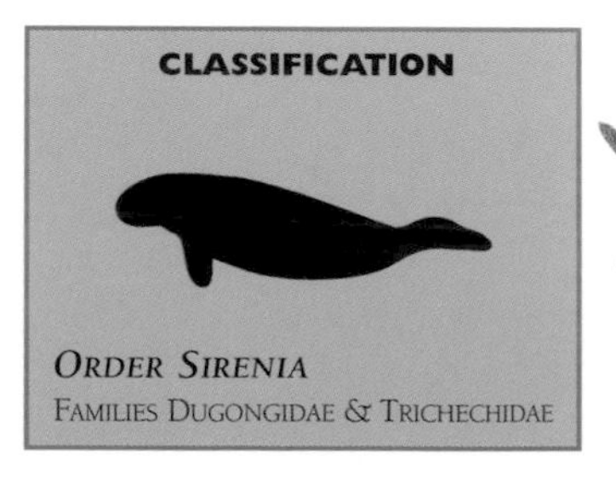

Order Sirenia
Families Dugongidae & Trichechidae

DUGONG DUGON

Dugong meat has a taste like that of domestic land mammals rather than fish, and has been successfully exploited by indigenous peoples for thousands of years. Aboriginal hunters name the large, old males "whistler" dugongs—they appear to use a whistling sound to keep their herds together. Bird-like chirps and high-pitched squeaks and squeals are used by females to communicate with their calves.

Sirenian distribution

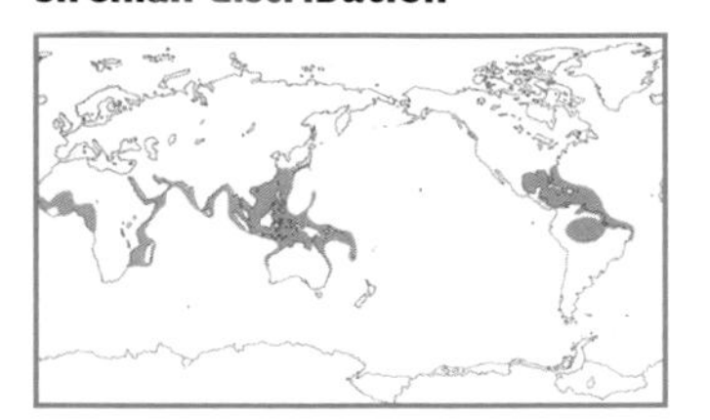

The four species of sea cows inhabit largely separate geographical ranges, restricted to the tropics and subtropics.

head, with its vast upper lip covered with stout sensory bristles, resembles that of a pig. Inside the downturned mouth are horny pads like those a cow uses for grasping grass. The most obvious difference in the manatees is the horizontal, paddle-shaped tail, like that of a beaver or platypus. West Indian and West African manatees weigh over 1,100 pounds (500 kg), while Amazonian manatees and dugongs average 660 pounds (300 kg) in weight.

TRICHECHUS SP.
The characteristic flattened, expanded tails of manatees are an example of convergent evolution with whales in response to adaptation to a marine environment, rather than shared ancestry.

Food The angle of the dugong's snout is more pronounced than that of the manatee and with its dense, heavy skeleton that keeps it on the bottom while feeding, the dugong is virtually obliged to feed on bottom-dwelling organisms. Manatees can also feed on plants growing at or near the surface, giving them an ecological edge over the dugong.

Young A female dugong is at least 10 years old before she bears her first calf, which is suckled for up to 18 months. Subsequent births are three to five years apart. The reproductive rate of manatees may be slightly higher.

Habitat Dugongs are restricted to the sea in the waters of 43 countries in the Indo-Pacific region. Manatees occur mainly in rivers and estuaries. The West Indian and West African manatees occupy similar freshwater and marine habitats on either side of the Atlantic. Amazonian manatees are restricted to freshwater habitats in the Amazon basin.

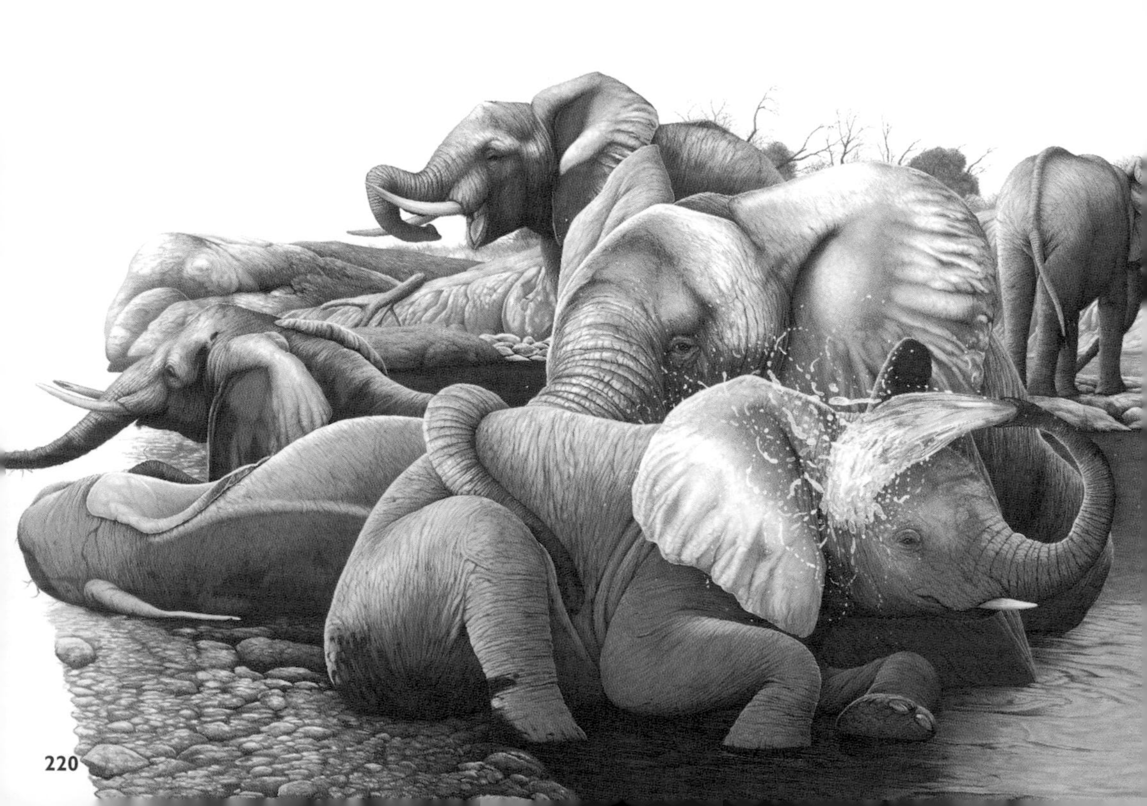

Kinds of Mammals

Elephants

Elephants are classified in the order Proboscidea, after their most distinguishing feature—the proboscis or trunk. The earliest proboscideans were pig-sized mammals without trunks, which lived 50–60 mya. With the exception of Antarctica, Australia and oceanic islands, they inhabited every continent on Earth, in every environmental extreme. Around 350 proboscidean species have been identified; today, two remain. These are the two genera of living elephants. The larger African elephant has two distinct subspecies. The more highly evolved Asian elephant has three subspecies that differ less obviously, but by a gradual change in characteristics across their range.

KINDS OF ELEPHANTS

The Asian elephant is more highly evolved and specialized than the African elephant. The Asian is closer in evolutionary terms to extinct mammoths than to the living African species. Consequently, differences within the Asian subspecies are not as distinct as differences among the African elephants. The forest African elephant is smaller than its bush African relative.

Similarities The most obvious shared similarity between the two elephant species is the possession of the trunk. Other shared external features are: the large size, with an overall arch-shaped body from the tip of the trunk to the tip of the tail; large ears; long tusks; lack of body hair; columnar legs; and the inguinal (between the hindlegs) position of the genitalia in both sexes. They also have a thick skin and many unique internal anatomical similarities, including the light, honeycomb bone structure of the skull.

Contrasts African elephants are generally taller and heavier and have larger ears than the Asians. The African has a concave back, as compared with the Asian, which is convex. Tusks are usually present in the African elephant of either sex; in the Asians, only in the males. The trunk of the African has more folds of skin in forms of "rings" and its tip possesses two, instead of one finger-like process. The trunk of the African species appears "floppy," while that of the Asian seems slightly more rigid.

AFRICAN ELEPHANT
Anatomical differences between the species are usually manifested in corresponding observable behavior. An African elephant picks up an object with the two "fingers" of its trunk to bring it to the mouth; while the Asian elephant curls its trunk around an object, which is "squeezed" and brought to the mouth.

ASIAN ELEPHANT

While the size of the ears is the most obvious superficial difference between the two elephant species, an important anatomical difference is the structure of the chewing surfaces of the teeth. Those on the tooth of an Asian elephant are narrow, compressed loops, but the African elephant tooth is wider and diamond-shaped. Asian elephants are grazers, while African elephants feed on tree foliage, bark and fruit.

Proboscidean distribution

The African elephant is distributed from Senegal in West Africa, across Central, East and southern Africa. The Asian elephant occurs in southern India and from Nepal across Indochina, and in Sumatra and Sri Lanka.

SUN PROTECTION

African elephants often appear brown from wallowing in mudholes of colored soil that becomes plastered on their bodies, although both species of elephant are usually gray in color. In Asia, wallowing usually results in a darker or lighter gray than the original gray body color.

African Elephant

This is the largest living land mammal. The elephant's trunk is its single most important feature, used for feeding, watering, dusting, smelling, touching, lifting, in sound production and communication, and as a weapon of defense and offense. The African elephant's ears are enormous and may reach half the height of an individual. They are around twice the surface area of the Asian elephant's ears.

Close relatives The genus *Loxodonta* contains two subspecies.

Characteristics The African elephant is immediately identifiable by the two "fingers" at the tip of the trunk, which itself can weigh 330–440 pounds (150–200 kg). An adult bush African male may reach 13 feet (4 m) in height and weigh 15,500 pounds (7,000 kg). The forest subspecies reaches 10,000 pounds (4,500 kg) and 10 feet (3 m) in height. About three-quarters of the elephant's life is spent feeding or moving toward a food or water source. Most elephants consume 165–330 pounds (75–150 kg) of food and 20–40 gallons (80–160 l) of water per day.

Food The diet is strictly herbivorous and the African elephant is adapted to be a browser, rather than grazer, on foliage, fruit and grasses.

Young The gestation period of 18–22 months is the longest of any mammal. The single offspring (rarely twins) weighs about 165–255

LOXODONTA AFRICANA AFRICANA
While the distribution of the bush African elephant and its forest-dwelling relative may overlap, the bush species occupies a variety of habitats, including arid savannas or deserts, to wet areas of marshes and lake shores, from sea level to mountain regions above the snowlines. The forest subspecies prefers equatorial forested regions, perhaps venturing into areas intermediate between forests and grasslands.

pounds (75–115 kg). The infant suckles for eight to 10 months and a female may give birth to about seven offspring in her 60–70 year lifespan.

Habitat The bush subspecies occupies a wider habitat range than the more specialized forest-dweller, but their overlapping distribution spans most of the African subcontinent below the Sahara Desert.

LOXODONTA AFRICANA CYCLOTIS

The forest African elephant is smaller with rounder ears and straighter and more slender tusks than the bush subspecies. It is considered the more primitive of the two subspecies, because the trend in mammalian evolution (except in island populations) is to become bigger. The original population probably lived in forested areas and contained smaller individuals. Subsequent generations grew bigger and taller and inhabited more open gallery forests and savannas.

CLASSIFICATION

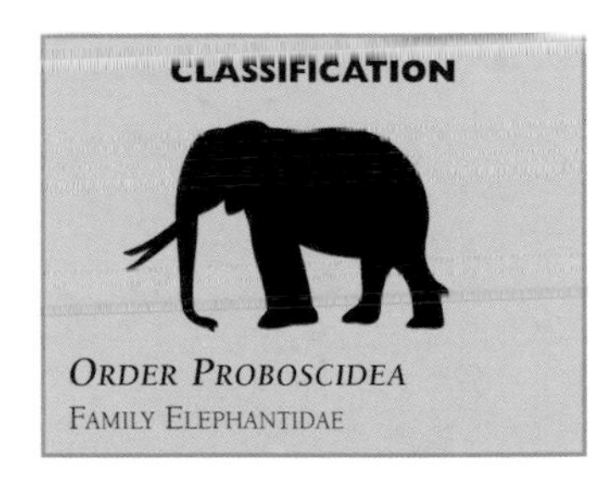

ORDER PROBOSCIDEA

FAMILY ELEPHANTIDAE

ASIAN ELEPHANT

The profile of the back of the Asian elephant, unlike that of the African elephant, is convex, or level, rather than concave. Its ears do not exceed the height of the neck. The Asian elephant has more hair on its body than the African species and it is especially conspicuous on the newborn and young.

Close relatives A very generalized gradual change in characteristics among the three Asian subspecies occurs across their range.

Characteristics The apparently more rigid trunk of the Asian elephant has only one "finger" at the tip. The trunk can weigh 275–440 pounds (125–200 kg), and the largest Sri Lankan male may reach over 11 feet (3.5 m) in height and weigh 12,000 pounds (5,500 kg).

Food The Asian elephant is adapted to be a grazer rather than a browser and a variety of grass species is consumed, as well as juicy leaves and fruit.

Young Like the African elephant, the Asian elephant secretes fluid from the musth gland, which is located beneath the skin, midway between the eye and the ear on each

ELEPHAS MAXIMUS INDICUS
The mainland Asian elephant (below), from the Indian subcontinent, across Indochina to China, exhibits a mixture of characteristics between the extremes of the easternmost and westernmost subspecies. Its size, color and degree of depigmentation are between the other two forms.

ELEPHAS MAXIMUS MAXIMUS
The Sri Lankan Asian elephant (above), found on the island of Sri Lanka and on the Indian subcontinent, is the largest of the three subspecies, is darkest in color and has the largest ears. It has distinct patches of depigmentation.

side of the head. It is not found in any other living mammal. Musth is a periodical phenomenon associated with male sexual activity. Musth does not occur in females, although they do secrete fluid that differs in chemical composition. Elephants reach sexual maturity between eight and 13 years of age. Newborns measure about 3 feet (1 m) tall at the shoulder and details of their reproduction are the same as for the African elephant.

Habitat The Asian elephant occupies a variety of habitats in 14 Asian countries, but generally prefers forested and transitional zones

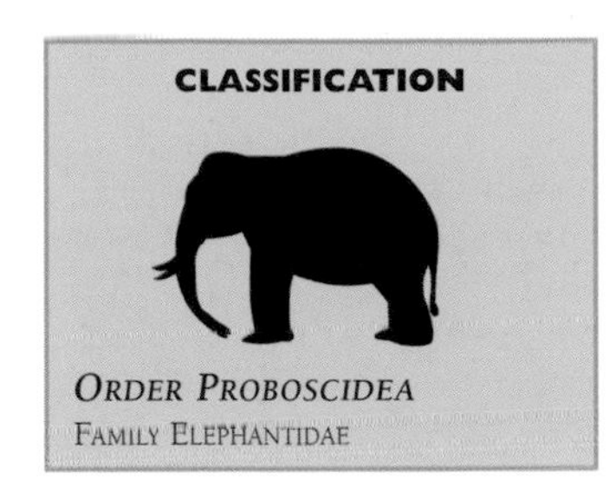

ELEPHAS MAXIMUS SUMATRANUS
The Sumatran Asian elephant (below), from Sumatra and southern Malaysia, is the smallest, the lightest in color and the least depigmented of the three subspecies.

ELEPHANT ANATOMY

The anatomy and physiology of an elephant are superbly adapted to its primary objective of keeping cool. An animal the size of an elephant has a small surface area compared with its body weight and cannot dissipate heat energy readily. Large ears with thin skin, an incredibly versatile trunk and many behavioral modifications solve this problem.

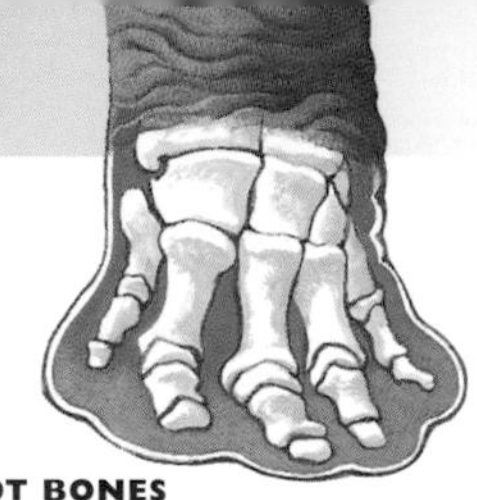

FOOT BONES
Despite their great weight, elephants walk almost on tiptoe. The digits are supported from behind by a fibrous fatty cushion and enclosed in a hoof-like structure of skin.

The ears Elephant ears function as cooling devices. Numerous blood vessels on the medial sides of the ears, where the skin is about one-twelfth of an inch (1–2 mm) thick, enable heat dissipation by ear flapping, particularly on warm days when there is little or no wind. The size of the ears is often related to climate: the African elephant, which originated closest to the equator and has remained within Africa, has the largest ears.

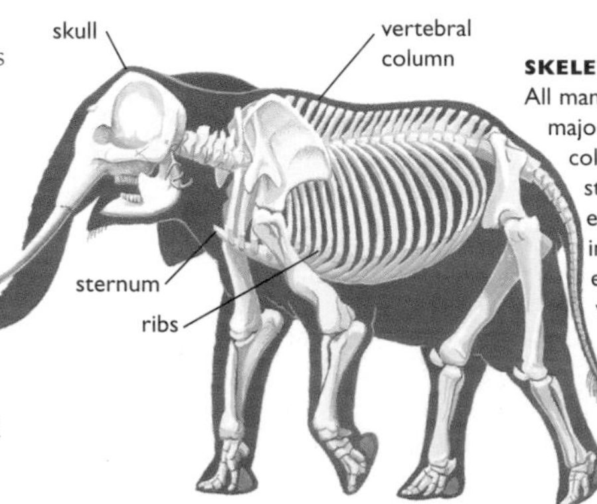

SKELETON OF AN ELEPHANT
All mammals have skeletons with four major divisions: the skull, vertebral column, appendages and ribs and sternum. Unlike other mammals, for example a dog, where the legs are in an angular position, the elephant's legs are in an almost vertical position under the body (similar to the legs of a table). This arrangement provides a strong support for the vertebral column, thoracic and abdominal contents and the great weight of the animal.

The skin The former name of the order, "Pachydermata," refers to the skin of the elephant. It is paper thin on the inside of the ears, around the mouth and the anus, and as thick as 1 inch (2.5 cm) or more around the back and in some places on the head. Even where thickest, the skin is a sensitive organ system with a rich nerve supply.

The trunk This is an organ of extraordinary complexity and flexibility. It is made up of at least 150,000 individual muscle units; is powerful enough to tear limbs from trees; and is so dextrous that it can easily pick up objects from the ground as small as a coin. It has no bones or cartilage. For cooling, the trunk can spray water, dust or mud to almost any part of the animal's body. It can hold more than 2 gallons (8 l) at a time.

ELEPHANT TUSKS

In elephants, milk incisors, or tusks, are replaced by permanent second incisors within six to 12 months of birth. Permanent tusks grow continuously at the rate of about 7 inches (17 cm) a year and are composed mostly of dentine. Like all mammalian teeth, elephant incisors have pulp cavities containing blood vessels and nerves; tusks are thus sensitive to external pressure. On average only about two-thirds of a tusk is visible externally, the rest being embedded in the socket within the cranium.

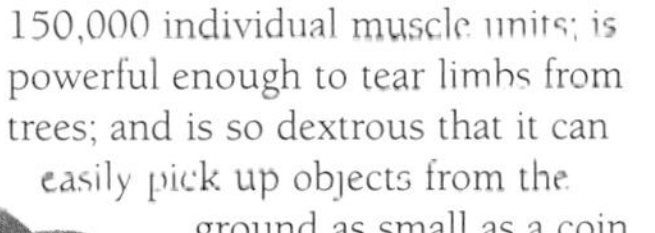

ELEPHANT AGING

There are several ways of estimating the age of elephants, from the technical—tusk circumference, body or eye lens weight, tooth eruption and wear—to the observable. The ear begins to curl between 20 and 40 and becomes torn at the edges with age. The face of a really old elephant is gaunt and sunken in appearance. Elephants have lifespans comparable with humans: about 70 years.

KINDS OF MAMMALS

UNGULATES

There are two orders of browsing or grazing hoofed mammals, called ungulates. The Perissodactyla are the odd-toed ungulates that support their body on the elongated third toe of each limb. They are the horses, asses and zebras; the tapir; and rhinoceros. In the horse family, the feet have a single (third) toe, which ends in a large hoof. The Artiodactyla, or even-toed ungulates, have two or four weight-bearing toes on each foot, ending in a hoof. The third and fourth toes are equally developed. They are the pigs; peccaries; hippopotamus; mouse, musk and antlered deer; giraffes; pronghorn; and numerous bovids. The orders are also differentiated by their mechanisms for digesting cellulose.

ODD AND EVEN TOES

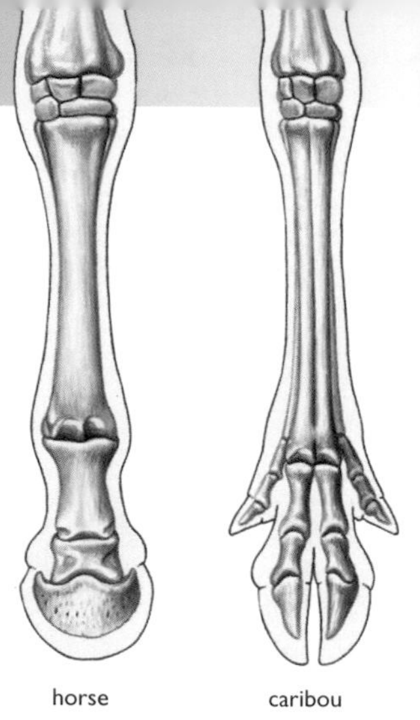

About 60 mya, when plant-eating mammals began to take advantage of open grasslands, they found that the only way they could escape from predators was to run. Because it was easier to run on their toes than with a flatter foot, their claws evolved into hard hooves and toes that were not needed for support disappeared or became smaller.

Specialized grazers Horses were the first ungulates to adapt to running on the tips of their toes across grassy plains. Tapirs have remained forest animals. Rhinos were the dominant form of large herbivore between 25 and 40 mya. Like these odd-toed ungulates, the biology of the even-toed, or cloven-hoofed, ungulates is dominated by means of evading predators. The bones of the soles of their feet are large, to absorb the stress of running. The hooves ensure immediate traction on almost any earth surface.

Ungulate distribution

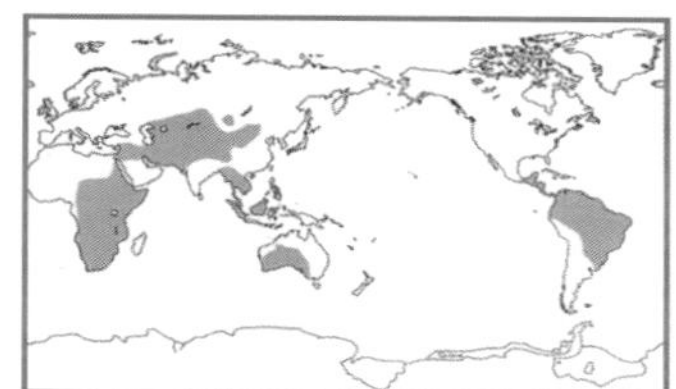

The map shows the distribution of odd-toed ungulates, which occur naturally in grasslands in Africa, Asia and South America. Even-toed ungulates have a wide geographic distribution in all continents except Australia and Antarctica.

ODDS AND EVENS

All hoofed mammals walk and run on their toes. Odd-toed ungulates (like the horse) have a middle toe larger than the other toes. Perissodactyls have either one or three toes. Even-toed ungulates (for example, the caribou) walk on their middle two toes. Artiodactyls have either four or only two toes.

ZEBRA FOOT
Zebras, like horses, run on the central toe of each foot, which ends in a hoof. The other toes are only stumps of bone. The other odd-toed ungulates—the tapir and rhinoceroses—have three toes.

RHINO FOOT
White rhinoceros have three toes on the front foot. The first and fifth toes have disappeared.

CAMEL FOOT
Camels walk on the third and fourth toes of each foot. The other toes have disappeared. In this respect they resemble deer and cattle, but camels do not support themselves on the tips of their toes. The toes are nearly horizontal and rest on fleshy cushions, with the hooves only serving to protect the front of the foot.

CARIBOU FOOT
Caribou have four toes, which can spread to provide support on soft snow. All artiodactyls, but the camels (and the hindfeet of peccaries), have four toes, even though in most, only the middle two are functional.

ZEBRA

The members of the older order of ungulates—perissodactyls—are alike in being odd-toed and able to digest the cellulose content of plant cell walls. The ancestors of all equines were probably striped, but there are now three species of zebras—the plains or Burchell's zebra, Grevy's zebra and the mountain zebra. Only the first is common; the others are endangered.

Close relatives On fossil and anatomical evidence, Grevy's zebra and the mountain zebra are as closely related to each other as they are to horses and asses.

Characteristics Equids' eyes are set far back in the skull and they have a wide field of view. They have binocular vision in front and a blind spot directly behind the head. Zebras have a keen sense of smell and can rotate their ears to locate sounds.

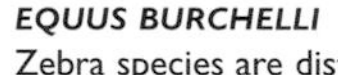

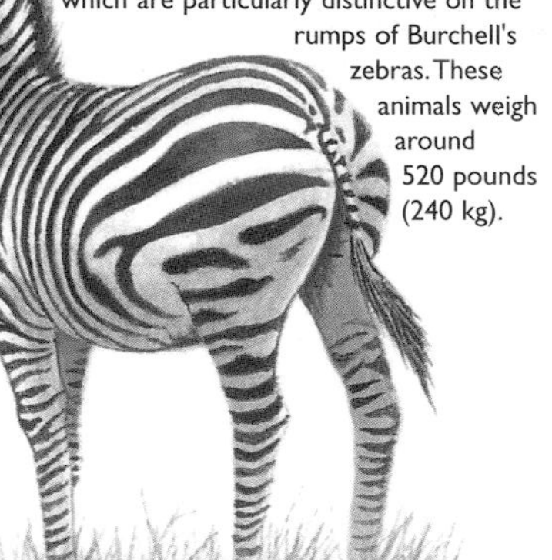

EQUUS BURCHELLI
Zebra species are distinguished from one another by the pattern of their stripes, which are particularly distinctive on the rumps of Burchell's zebras. These animals weigh around 520 pounds (240 kg).

SAFETY IN NUMBERS
Zebras must drink regularly. They are social animals and the daily line-up at waterholes is a familiar sight on the African plains. Zebras do not have much stamina and can easily be run down when pursued on horseback.

Food Zebras are nomadic grazers. Leaves and bark sometimes form part of the diet but they eat the coarser parts of grasses and so do not compete strongly for food with the various grazing antelope that share the same territory.

Young Males use the flehmen or lip-curl response to assess the sexual state of females and compete fiercely for receptive ones. The female bears a single young, rarely twins, after a gestation period of about a year. Foals are up and about within an hour of birth. Mothers whinny when separated from their foals and nicker to warn them of danger.

Habitat Burchell's zebras are distributed across the grasslands and savanna of East Africa.

CLASSIFICATION

ORDER PERISSODACTYLA

FAMILY EQUIDAE

ON THE ATTACK

Adult zebras are hunted by lions; hyenas and wild dogs prey on foals. Stallions actively try to protect their young from predators by kicking out with their hindlegs.

Burro

The burro, donkey or domestic ass, a direct descendant of the African wild ass, has been used as a beast of burden since 4000 BC. In many parts of the world, feral populations have grown from escaped or abandoned animals. In the desert areas of western USA, feral burros compete with bighorn sheep for scarce food.

Close relatives The burro resembles other members of the horse family.

Characteristics Different breeds of burros vary greatly in size. Donkeys in Sicily are only about 24 inches (60 cm) tall while the American ass may exceed 63 inches (160 cm). Burros may be white, gray, brown or black, usually marked with darker stripes from mane to tail and across the shoulders.

Food Feral burros graze on available plant material. They can survive in such hostile environments as Death Valley in south-eastern California.

EQUUS ASINUS
Burros cannot move as fast as horses but they are surefooted pack animals and capable of carrying heavy loads in rough and steep terrain. The donkey's reputation ranges from sacred animal and sturdy, faithful beast to the definitive symbol for stubborn stupidity.

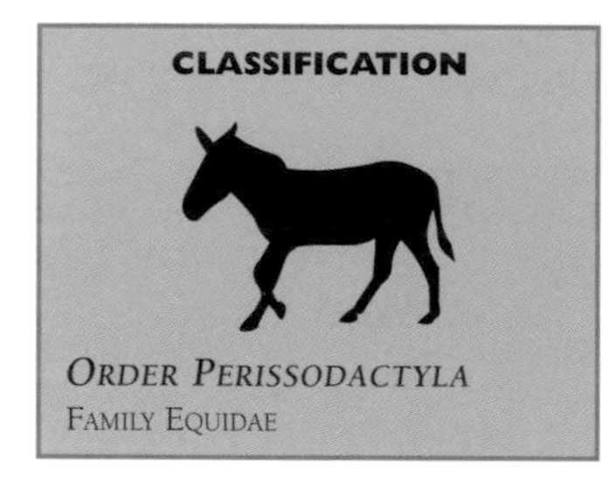

Young Burros usually give birth to single offspring. The mating of a male burro and a female horse produces a hybrid mule.

Habitat Burros are distributed widely across Mediterranean Europe, the Near East, northern Asia, America and much of Africa. They thrive in domesticated conditions in Britain and in feral herds in the arid center and north of Australia.

Brazilian Tapir

The tapir's way of life has changed little since ancestral forms existed 20 million years ago. Tapirs are usually solitary, except for mothers with their young.

Close relatives There are three South American species of tapirs and a single Asian species.

Characteristics Tapirs are squat animals weighing about 660 pounds (300 kg). Their head and body length is 70–100 inches (180–250 cm).

TAPIRUS TERRESTRIS

These long-surviving mammals face an uncertain future. Their habitat is threatened by land clearing and the flooding of forests by dam developments; they are hunted extensively for food, sport and their thick skins. The latter provides good-quality leather that is made into whips and bridles.

CLASSIFICATION

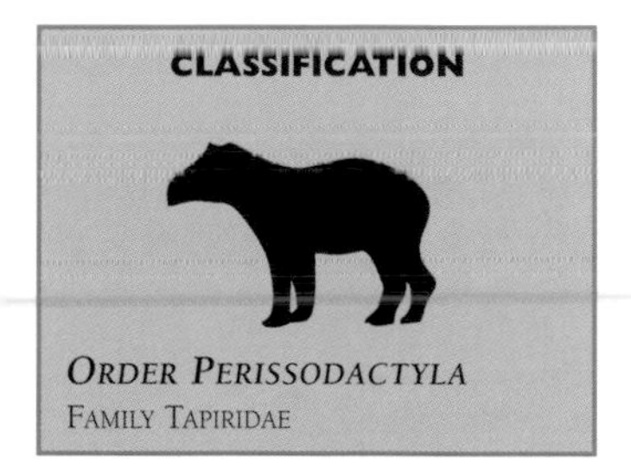

ORDER PERISSODACTYLA
FAMILY TAPIRIDAE

They are excellent swimmers and divers, can move quickly on land and are able to climb well.

Food The Brazilian tapir uses its mobile snout to strip the parts from trees, fruit, grasses and water plants.

Young Pregnant females give birth to one, occasionally two, offspring in a secure lair. Newborn tapir are born with pale flecks and stripes, which helps them to blend with their environment.

Habitat The Brazilian tapir occupies rainforest near water or swamps in South America from Colombia and Venezuela, south to Brazil and Paraguay.

Rhinoceros

There are two species of two-horned African rhinoceros: the black or hook-lipped rhino and the white or square-lipped rhino. They evolved from the same stock during the course of the Pliocene epoch (about 3 mya). In some parts of the world rhino horns fetch an extremely high price and are ground into powder and taken as medicine or carved into dagger handles.

Close relatives The five surviving species of rhinoceros fall into three distinct subfamilies: one restricted to Africa; the other two to Asia.

Characteristics Rhinos have poor vision, a good sense of hearing and an excellent sense of smell. They vary in weight from the Sumatran rhino up to 1,750 pounds (800 kg) to the massive bulk of the male African white rhino, which can reach 5,000 pounds (2,200 kg).

Food The black rhino browses, mostly at night, using its prehensile upper lip to draw branch tips into its mouth. The white rhino grazes, cropping short grasses with its wide lips.

Young Black rhino calves have a 15-month gestation; white rhino

DICEROS BICORNIS

The black rhino is actually gray and not greatly different in color from its white relative. The precise shade of a rhino's hide is determined by the color of the local soil in which it rolls to keep cool and rid itself of insect pests.

DEFENSE FORMATION

Rhinos are basically solitary animals but white rhinos are the most sociable of the five species. Females with calves sometimes team up in small groups. If predators threaten their calves, they form a circle to protect them.

Javan rhinoceros height: 72 in (180 cm); males have a single horn up to 11 in (28 cm); heavily folded and patterned skin.

Indian rhinoceros height: 73 in (182 cm); single horn up to 24 in (60 cm) folded skin studded with bony nodules.

Sumatran rhinoceros height: 53 in (132 cm); two short horns; folded skin covered in red to black bristles.

White rhinoceros height: 79 in (198 cm); two horns—front horn up to 63 in (157 cm) long; smooth skin.

Black rhinoceros height: 61 in (152 cm); two horns—front horn up to 54 in (135 cm) long; smooth skin.

calves a month longer. Births peak from the end of the rainy season through to the middle of the dry.

Habitat The black rhino is found from the Cape to Somalia, in thick bush habitats. The white lives in southern and north-eastern Africa, in different kinds of savanna.

CLASSIFICATION

ORDER PERISSODACTYLA
FAMILY RHINOCEROTIDAE

CERATOTHERIUM SIMUM
Fossil remains indicate that the white rhino was once more widely distributed through East and North Africa. Hunted almost to extinction, its numbers have been increasing under careful protection.

Chacoan Peccary

The Artiodactyla far outnumber the odd-toed ungulates in diversity of species. Artiodactyls are herbivores and are widely distributed across all continents except Australia and Antarctica. Peccaries, pig-like in form, have more complex stomachs than pigs, indicating a longer evolutionary period as herbivores.

Close relatives This species is related to the collared and white-lipped peccaries.

CLASSIFICATION

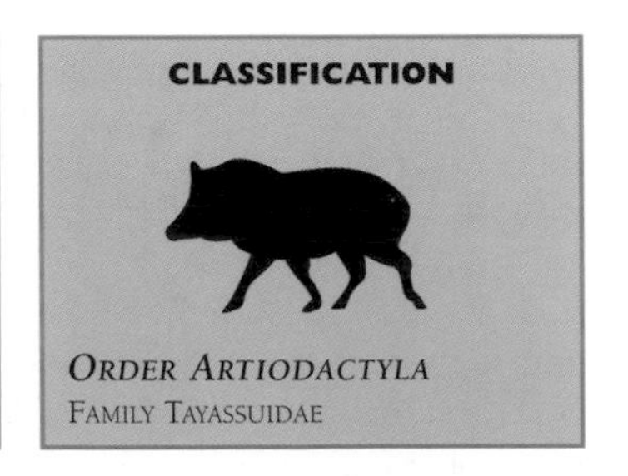

Order Artiodactyla
Family Tayassuidae

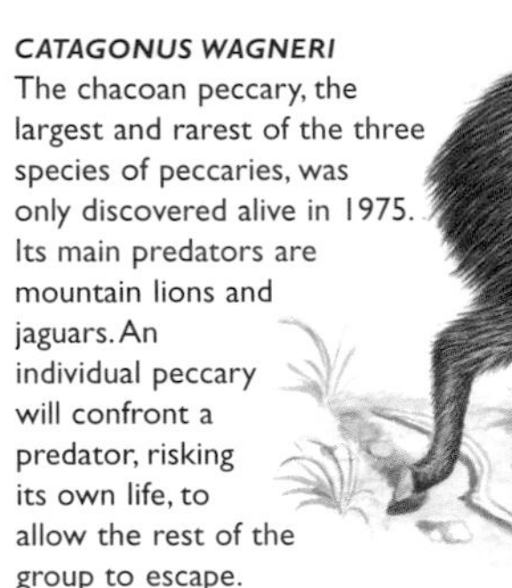

CATAGONUS WAGNERI
The chacoan peccary, the largest and rarest of the three species of peccaries, was only discovered alive in 1975. Its main predators are mountain lions and jaguars. An individual peccary will confront a predator, risking its own life, to allow the rest of the group to escape.

Characteristics The fused cannon bones (the greatly developed middle foot bone) in peccary hindlegs reveal a history of sprinting from predators over soft ground. They live in close-knit groups and jointly defend their territories. Males and females, alike in external appearance and size, are fighters and have sharp canines. Chacoan peccaries range in head and body length from 40 to 45 inches (102–120 cm). They weigh between 70 and 95 pounds (30–45 kg).

Food During the daytime, peccaries find much of their food below ground as roots and tubers.

Young Peccaries bear few and very small young. These are active soon after birth and exposed immediately to their environment.

Habitat Chacoan peccaries inhabit the thorny forest and isolated areas of palm and grasses in South America.

BABIRUSA

The family Suidae is a diverse group that evolved in Africa, a giant species of pig appearing early in the ice ages. Large tusks are generally used in fighting as a means of holding and controlling the opponent's head, but the babirusa does not conform in this respect.

Close relatives The family Suidae is divided into nine species in five genera.

Characteristics The babirusa is a large, pig-like mammal and in the male the tusks curve upward over the eyes. It has four toes on each foot, sparse, short white or gray bristles, and no facial warts. The head and body length is between 33 and 40 inches (85–105 cm). A babirusa can weigh up to 200 pounds (90 kg).

Food The babirusa feeds largely on fruit and grass.

Young One or two young are born in a litter and the piglets occupy a grass nest for about 10 days until they are strong enough to follow their mother. Females have two mammae, one for each offspring.

Habitat The babirusa is found only in the tropical forest of Sulawesi and some nearby islands.

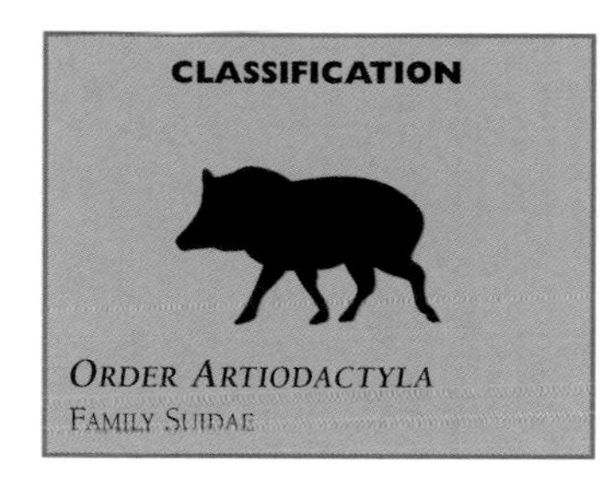

CLASSIFICATION

ORDER ARTIODACTYLA
FAMILY SUIDAE

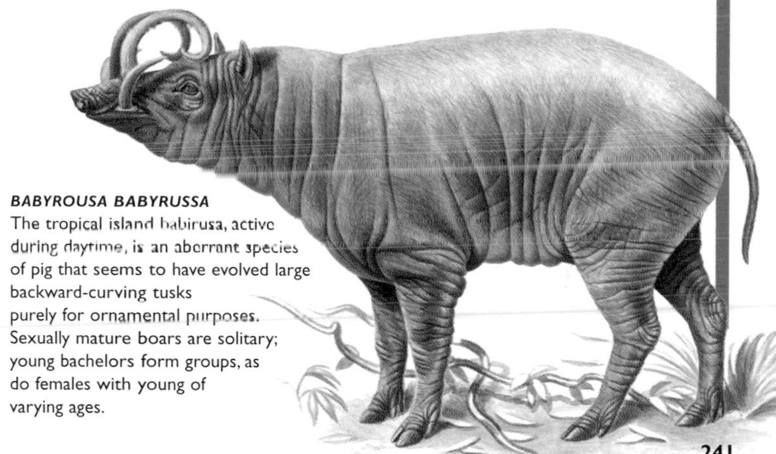

BABYROUSA BABYRUSSA
The tropical island babirusa, active during daytime, is an aberrant species of pig that seems to have evolved large backward-curving tusks purely for ornamental purposes. Sexually mature boars are solitary; young bachelors form groups, as do females with young of varying ages.

Red River Hog

Also known as the bush pig, the red river hog is active day and night. It is a symptomless carrier of African swine fever, a virus disease transmitted by soft-bodied ticks that is lethal to domestic swine.

Close relatives All members of the family, including giant forest hogs and warthogs, are relatives.

Characteristics In common with other artiodactyls, red river hogs can run rapidly and have specialized teeth for cutting and grinding vegetation. Wild pigs have well-developed, upturned canines and molars with rounded cusps. Their noses are supported by a prenasal bone. Red river hogs range in size from 40 to 60 inches (105–150 cm), in weight from 110 to 260 pounds (50–120 kg).

Food The red river hog travels large distances in its search for food and will eat almost anything, including grass, roots, fruit and grain crops, small mammals, birds and carrion in its diet.

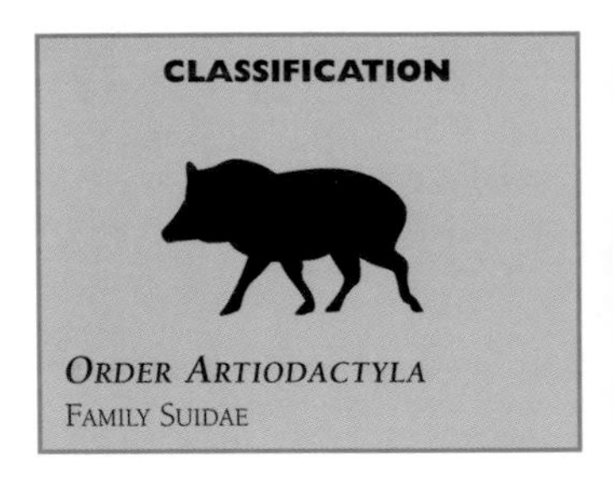

Young The female, which has three pairs of mammae, makes a grassy nest and gives birth to three to six young. The piglets' coats are striped at birth.

Habitat Red river hogs are distributed south of the Sahara and on the island of Madagascar, inhabiting terrain where they can find cover.

POTAMOCHOERUS PORCUS
Distinctive among pigs for its white dorsal mane and whiskers, the red river hog has several color forms, not all of them red. Males have facial warts. These gregarious animals live in groups of around 12, led by an old male.

Wild Boar

The wild boar has been domesticated for food in many cultures since early Neolithic times. The sociable disposition of pigs and the lack of territorial defense makes them ideal for domestication. Wild boar males have short, sharp tusks for attack and thick dermal shields across the shoulders for defense.

SUS SCROFA
The young of the wild form are born with striped coats but feral piglets are plain colored. Females may bear several litters a year and shelter their numerous offspring from inclement weather in domed nests of grass.

Close relatives Wild boars are related to all other members of the Suidae family.

Characteristics As with all pigs, the males are considerably larger than the females. Head and body length is 35–70 inches (91–180 cm); weight is 110–440 pounds (50–200 kg).

Food Pigs are the most omnivorous of the cloven-hoofed mammals and root for food in moist soil. Wild boars search for food in the daytime and at twilight. They dig for bulbs and tubers and eat nuts and other plant material, insect larvae and carrion.

Young Female wild boars have six pairs of mammae and bear litters of up to 10 young after a gestation period of 115 days.

Habitat This species is widely distributed through Europe, North Africa, Asia, Sumatra, Japan and Taiwan. It has been introduced into North America and feral domestic populations, also called wild boar, occur in Australia, New Zealand and North and South America.

Hippopotamus

Fossil records show that during interglacials hippos were found as far north as England. A dwarf species colonized the Mediterranean islands. Females are fiercely protective of their young but hippos will rarely attack humans unless they are provoked or encountered on their marked paths.

HIPPOPOTAMUS AMPHIBIUS
Water loss from a hippo's skin in dry air is three to five times the rate in humans. Specialized skin glands all over its body secrete an oily pink substance. This protects the skin from sunburn and may also help to combat infection.

Close relatives The family Hippopotamidae consists of two species in two genera. Hippos are also closely related to pigs.

Characteristics The average length of the "true" hippopotamus is 11½ feet (3.5 m); its average weight is 5,300 pounds (2,400 kg). The pygmy hippo is only about a tenth of the mass of the "true" hippopotamus. *Hippopotamus amphibius* has four webbed toes on each foot; only the front toes of pygmy hippos are webbed. Hippos can close their nostrils and submerge for up to five minutes.

COMMUNAL WALLOWING
Both species of hippo conserve energy by resting in wallows during the daytime. At sunset they move out to a grazing area and feed for five to six hours. Each individual uses an established path, marked at intervals by piles of its dung. Hippos crop short grasses with their broad lips, an action that pulls up many varieties by the roots, leading to soil erosion.

Food Hippos are entirely herbivorous. They ferment grasses in a large, complex stomach.

Young Both species prefer to mate in water. The mother leaves her group to bear a single offspring

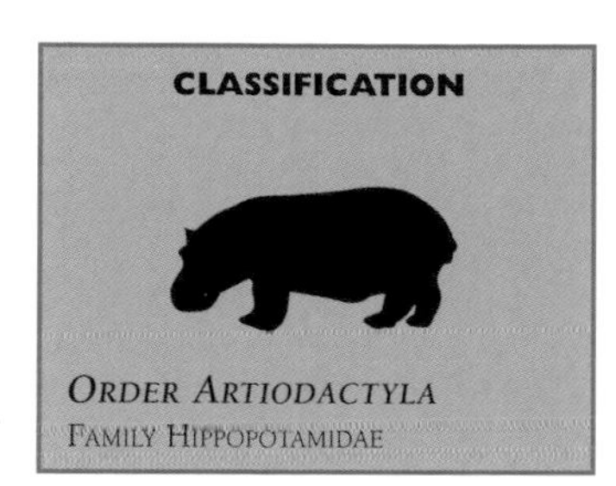

CLASSIFICATION

ORDER ARTIODACTYLA
FAMILY HIPPOPOTAMIDAE

(twins are extremely rare) on land or in shallow water. She attends to her newborn for about 10 to 14 days before rejoining the herd. Young hippos suckle for about eight months and ride piggyback on their mothers in deep water.

Habitat *Hippopotamus amphibius* occupies short grasslands, wallows, rivers and lakes south of the Sahara to Namibia and the Transvaal. Pygmy hippos, which are less aquatic, are found in rainforest, swamps and thickets near water from Guinea to Nigeria.

FRINGE BENEFITS

Hippos excrete vast quantities of fertilizing manure into lakes and rivers, thus encouraging the growth of plankton and fish. Cyprinid fish graze on algae on hippos' skin; oxpeckers help clear it of parasites. Terrapins and young crocodiles bask on hippos' backs; storks and egrets perch on them to fish.

FURIOUS FIGHTING

Bulls have exclusive rights in their territories, dominant males maintaining these for up to eight years. Fights between rivals may last as long as 90 minutes and are potentially lethal, as the lower tusks are long and sharp.

Bactrian Camel

Camels evolved in North America, developing in the Tertiary period from tiny gazelle camels to huge giraffe-like browsers. The Bactrian camel has two humps.

Close relatives The Bactrian camel's closest relative is the Arabian camel or dromedary, which has a single hump.

Characteristics All Camelidae have combat teeth, in which the canines and first premolars have been formed into caniforme fighting teeth. Camels are the only mammals to have red blood cells that contain a nucleus.

Food Camels eat dry, thorny, desert vegetation when more attractive food is unavailable. When the feed is good, they store fat in their humps, which they use as a reserve in lean times and for the manufacture of water by the oxidation of the fat. They have been known to go without drinking for up to 17 days.

CAMELUS BACTRIANUS
Camels see and smell remarkably well; they have double rows of heavy protective eyelashes, haired ear openings and the ability to close their nostrils. Their feet, though smaller than the feet of dromedaries, are admirably suited to walking on sand or snow. Bactrian camels lack the horny pads on the chest and knees that support the dromedary's weight when kneeling.

Young Female camels give birth to one large, well-developed young after a gestation period of 11 months and suckle it for a year.

Habitat Ninety percent of the world's camels are dromedaries. They are adapted to the dry deserts of northern Africa, Arabia and western Asia, but now the only free-living populations occur in Australia. The last of the wild Bactrian camels occupy the cold deserts of Mongolia. They are greatly endangered.

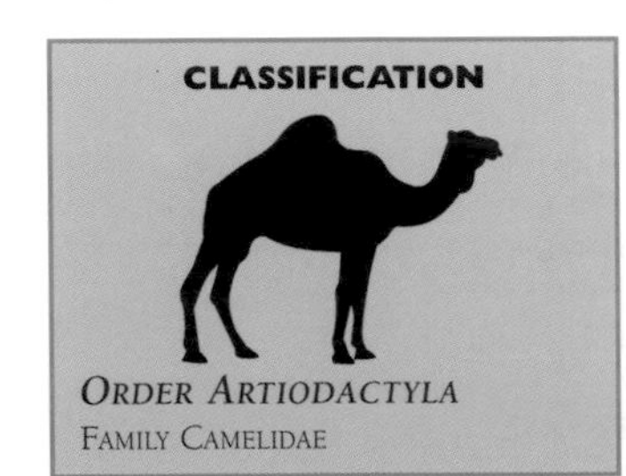

Llama

The four South American camel species (lamoids) are much smaller than the Old World camels and have no humps. They originated in North America and have occupied South America only since the early Pleistocene period (about 2–3 million years ago).

Close relatives The New World camel species—the alpaca, guanaco and vicuna—are most closely related.

Characteristics The llama has a long neck and legs and a relatively small head. It is the largest lamoid averaging a shoulder height of 47 inches (120 cm). Llamas are usually docile but will spit if annoyed. A 250-pound (113-kg) llama can carry a load of 100–132 pounds (45–60 kg) for an average of 15–20 miles (25–30 km) a day.

Food Llamas graze on grass and other plants. Like other camelids, they have a complex three-chambered stomach and ruminate or chew the cud. They have a high thirst tolerance.

Young The female gives birth to one young after a gestation period of nearly a year. Lamoids do not lick the newborn or eat the afterbirth. All four species interbreed readily with one another and produce fertile offspring, which are mobile at 15–30 minutes old.

Habitat Today, no llamas live independently of humans in their native Andean homeland. Herds are maintained by the Indians of Bolivia, Peru, Ecuador, Chile and Argentina.

LAMA GLAMA
Llamas are normally shorn every two years and each animal yields about 6–7 pounds (3–3.5 kg) of fiber. The fleece consist of a protective outer layer of coarse guard hairs that makes up about 20 percent of the coat and the short, crimped fibers of the insulating undercoat.

CLASSIFICATION

ORDER ARTIODACTYLA
FAMILY CAMELIDAE

Guanaco

The guanaco, which is twice the size of a vicuña, occurs in both sedentary and migratory wild populations. They are sociable animals. Males are polygamous and lead harems of 10 to 12 females with their young, which are called guanaquito. Young males and males without harems form herds, too.

Close relatives Guanacos' closest relatives are the three other New World camelid species: llama, alpaca and vicuña.

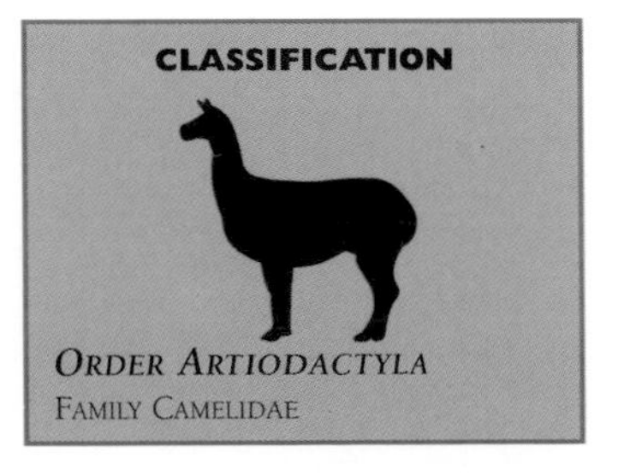

Characteristics The slender, long-limbed guanaco is able to move quickly over rugged terrain. The bodyweight of camelids rests on the sole-pads and only the front ends of the hoofs touch the ground—the pads of lamoids' toes are not as wide as camels'. Guanacos have an average shoulder height of 43–45 inches (108–120 cm).

Food The guanaco is both a grazer and a browser, and can find food in sparsely vegetated open country. It does not need to drink.

Young The female produces a single young every other year after a gestation period of 10 to 11 months. The adult male forcibly expels yearlings from the family group.

Habitat Guanacos occupy desert, grassland, savanna, shrubland and occasional forest. They range from the snowline to sea level through the Andes from Peru and Bolivia southward to Tierra del Fuego and other islands.

LAMA GUANICÕE

Guanaco pelts resemble those of red foxes in texture. The soft, downy, guanaquito fiber is especially prized for use in luxury fabrics. The guanaco is protected in Chile and Peru but not in Argentina, which exports tens of thousands of pelts annually.

WATER CHEVROTAIN

Regarded as living fossils, chevrotains have remained virtually unchanged in 30 million years of evolution and are intermediate in form between pigs and deer. Except when mating, they are solitary animals.

Close relatives There are four species of tropical tragulids: the three mouse deer in Asia and the water chevrotain of Africa.

Characteristics Water chevrotains have no cannon bones in their front legs. Their premolars still have conical crowns, much like those of the primitive ungulates from the early tertiary. Females may be bigger than the males and average a head and body length of 30 inches (77 cm), a shoulder height of 14 inches (35 cm) and weigh 22 pounds (10 kg).

Food A night forager, water chevrotain feed on plants and fallen fruit but also eat some insects, crabs, fish, worms and small mammals.

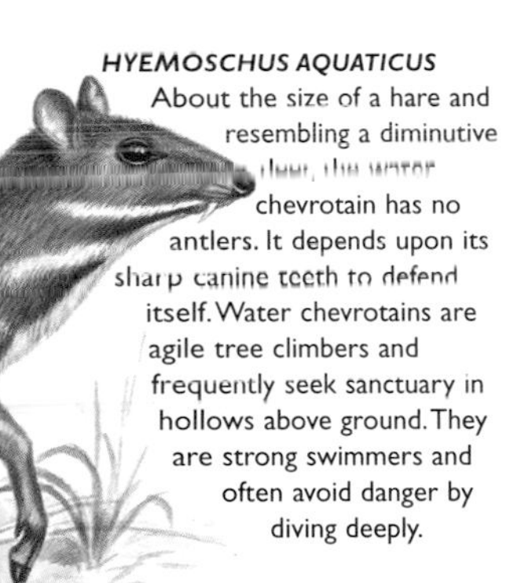

HYEMOSCHUS AQUATICUS
About the size of a hare and resembling a diminutive deer, the water chevrotain has no antlers. It depends upon its sharp canine teeth to defend itself. Water chevrotains are agile tree climbers and frequently seek sanctuary in hollows above ground. They are strong swimmers and often avoid danger by diving deeply.

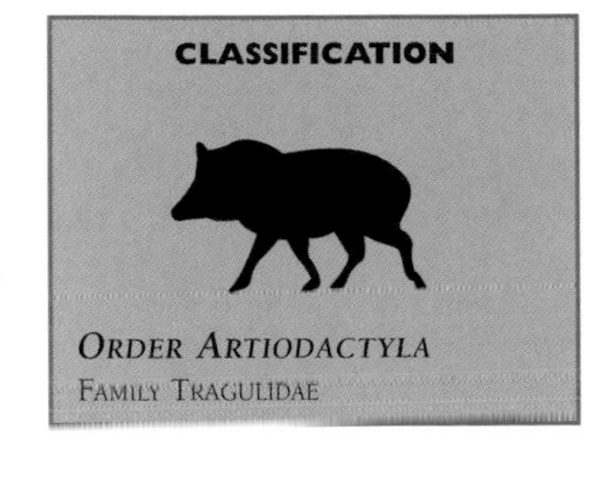

CLASSIFICATION

ORDER ARTIODACTYLA
FAMILY TRAGULIDAE

Young Males and females are attracted to one another by scent, and mate without aggression. The female is pregnant for about four months and produces a single offspring. Young suckle for several months, but eat some solid food after 14 days.

Habitat Water chevrotains are found from Guinea to Cameroon, Zaire, Gabon and the Central African Republic. They live in forests, always near water.

Pudu

Deer have long fascinated humankind and were painted on the walls of caves as much as 14,000 years ago. They are advanced ruminants and all species eat mainly woody vegetation, with the exception of the caribou. Bone antlers, in males only—except again, for caribou—are shed and regrown annually.

Close relatives The southern pudu shares the subfamily Odocoilinae (hollow-toothed deer) with some much larger deer, namely moose and reindeer. Their closest relative is the northern pudu.

Characteristics The pudu is the smallest species of the large family of true deer with a shoulder height of 14–15 inches (35–38 cm). It has simple unbranched antlers. This animal weighs only about 1 percent of an average moose's 1,763 pounds (800 kg).

PUDU PUDA
Like the less common northern pudu, which inhabits the forests and swampy, elevated savanna of Colombia to north Peru, the southern pudu's fur is thick and dense. Both species have a rounded back and short, delicate legs but the southern pudu is slightly smaller.

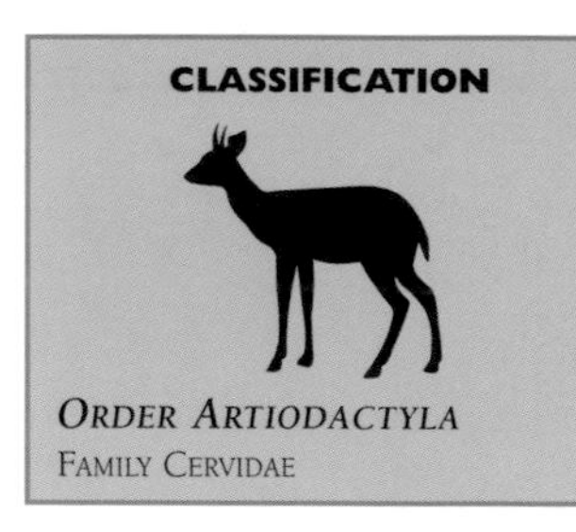

Food Southern pudus graze or browse for their food, eating grasses and the shoots, twigs, leaves, flowers and fruit of herbs, shrubs and trees.

Young Females give birth to single young, or sometimes twins. They are spotted when born.

Habitat Southern pudus hide in the deep forests clothing the Lower Andes of Chile and Argentina.

Indian Muntjac

Indian muntjacs are small as deer go. They are nocturnal and either solitary or move about in pairs or small family groups. They are more or less completely covered in hair but, like most other deer, have a small naked patch on the muzzle. They are also called barking deer on account of their cry.

Close relatives The genus comprises six species.

Characteristics In common with other Cervidae, females are recognizably smaller than males, which average a body weight of about 48 pounds (22 kg). Males have very simple antlers and both sexes have well-developed upper canine teeth. The coat is darker in winter than in summer.

Food The Indian muntjac's preferred diet is herbs, leaves, fruit, mushrooms and bark.

MUNTIACUS MUNTJAC
In common with most other species of deer, Indian muntjacs are preyed upon by humans and other animals. They are always on guard and use sight, hearing and smell to detect danger. If they hear or see movement, they flee rapidly.

Young One or two young are born after a gestation period of about 180 days.

Habitat Indian muntjacs inhabit thickly vegetated areas in which they can conceal themselves. They occur in the wild in India, Sri Lanka, Tibet, south-western China, Burma, Thailand, Vietnam, Malaya, Sumatra, Java and Borneo and have been introduced into England.

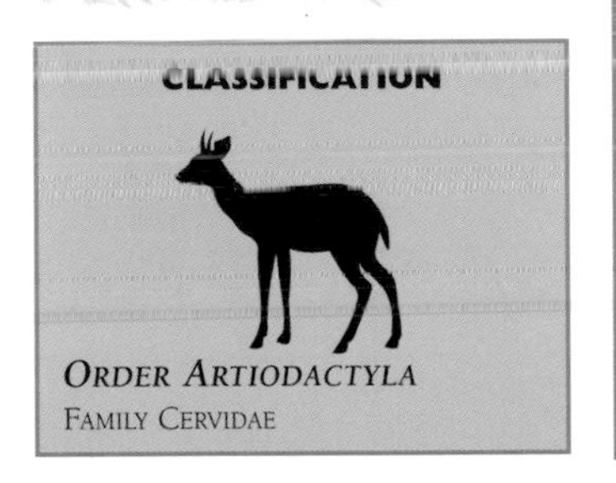

Fallow Deer

Fallow deer were widespread throughout Europe some 100,000 years ago, but most of the world's population now live in deer parks or reserves. The Norman kings of England planted forests in southern England to ensure an adequate supply of these animals for the hunt.

Close relatives Fallow deer belong to the subfamily Cervinae and are closely related to a number of other species, including red deer and wapiti.

Characteristics Male fallow deer may reach a weight of more than 220 pounds (100 kg) and have a prominent Adam's apple, as well as impressive branched and palmate antlers. Their coats are usually fawn with spots on the back and flanks in summer, which disappear in the winter when the hair turns grayish brown in color.

DAMA DAMA
Growing new headgear every year is an itchy business. Bucks shred the bark of trees in their attempts to rid themselves of old antlers and the remnants of protective skin, called velvet, which encases their new ones.

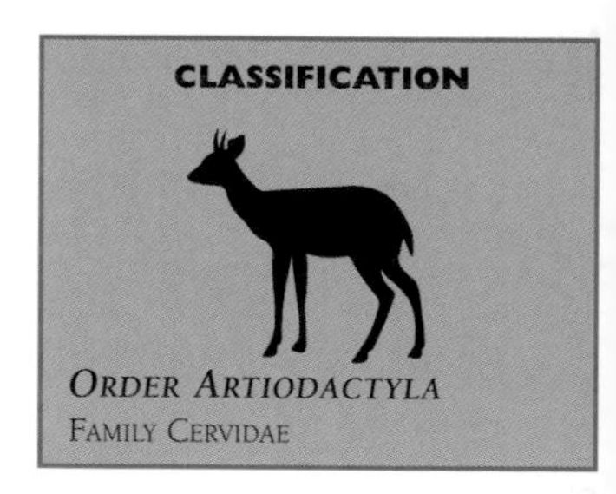

CLASSIFICATION

Order Artiodactyla
Family Cervidae

Food Fallow deer favor ground vegetation, especially beechnuts when available. They move to open space to graze at dawn and dusk and spend the rest of the day ruminating and sleeping.

Young The sexes remain separate for much of the year and mate in the fall. Young are almost always spotted.

Habitat Fallow deer are native to the woodland of Europe, Asia Minor and Iran. They have been introduced into Australia and New Zealand.

Wapiti

The wapiti is also called the American elk. During the rutting season wapiti make a high-pitched, whistling sound. Wapiti live in large herds in winter, breaking into smaller groups for the summer. Older bulls live alone or with two or three others.

Close relatives Members of the subfamily Cervinae, wapiti are related to red deer, chital, sambar, fallow deer and other species in this diverse group.

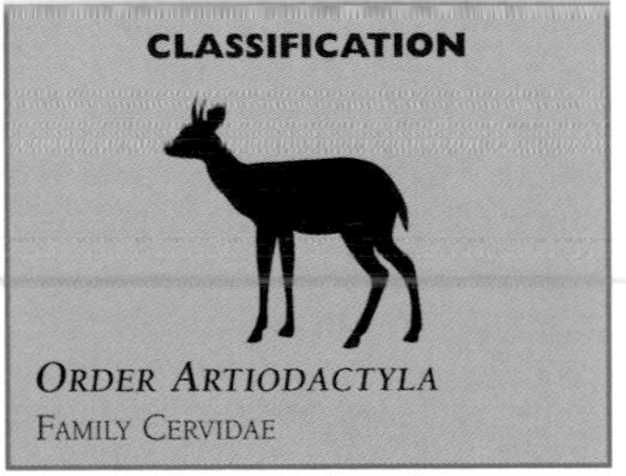

CLASSIFICATION

Order Artiodactyla
Family Cervidae

Characteristics Second in size to the moose, the male wapiti may stand taller than 5 feet (1.5 m) at the shoulder and weigh up to 990 pounds (450 kg). The coat is darker in the winter—an usual trait for an animal living seasonally in snow. The male's branched antlers tower above his head and usually bear five tines.

Food The wapiti consumes a huge daily intake of grasses and leaves and other woody foodstuffs. In severe winters, hungry animals may raid cultivated crops, orchards and haystacks.

Young Young wapiti are born with spots after a gestation period of 249–262 days.

Habitat Once common across North America, the wapiti now occurs naturally only in the Rocky Mountains, western US and Canada.

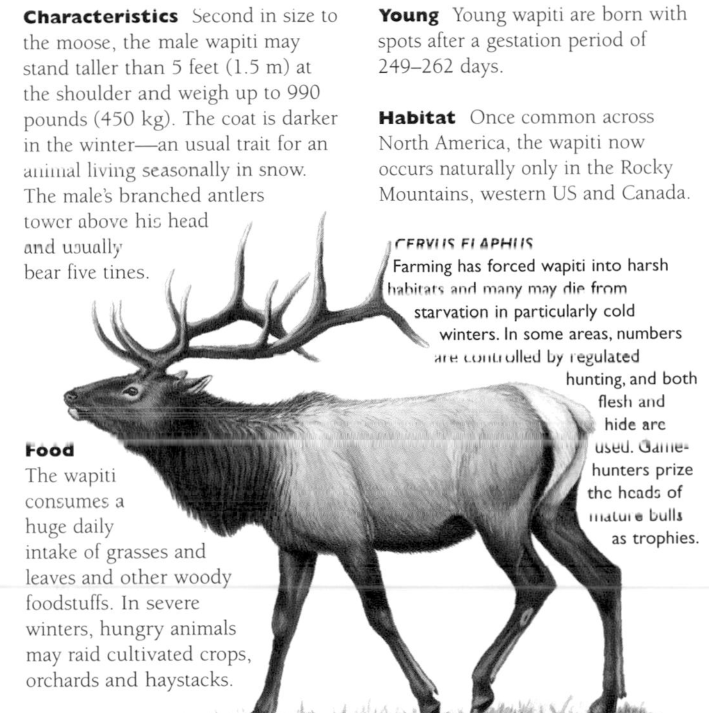

CERVUS ELAPHUS
Farming has forced wapiti into harsh habitats and many may die from starvation in particularly cold winters. In some areas, numbers are controlled by regulated hunting, and both flesh and hide are used. Game-hunters prize the heads of mature bulls as trophies.

Moose

Moose are the largest species of deer and are heavy, long-legged, short-necked ruminants. They are shy animals but are known to be unpredictable and aggressive. They are among the species of artiodactyls that specialize in obstructing a pursuer's path, trotting smoothly and fast over low obstacles when hunted by bears or wolves.

Close relatives Moose belong to the subfamily Odocoilinae or hollow-toothed deer.

Characteristics The shoulder height of a moose may be between 67 and 92 inches (168–235 cm). The naked patch on the pendulous muzzle is small. Moose move with a stiff-legged, shuffling walk that is surprisingly fast; they trot when fleeing, but rarely run.

Food When browsing on land, a moose consumes more than 9 pounds (4 kg) dry weight of terrestrial vegetation a day, or the equivalent of more than 20,000 leaves. These animals also have a taste for sodium-rich aquatic plants when they can get them.

ALCES ALCES

Moose are distinguished in appearance by humped shoulders and a fleshy dewlap (the "bell") of skin and hair up to 20 inches (50 cm) long that hangs down from the throat. Although generally protected, bull moose are hunted under strict controls for their huge antlers and heads.

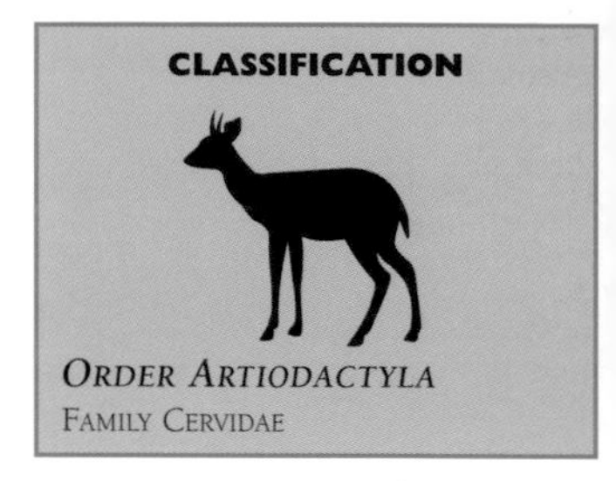

CLASSIFICATION

ORDER ARTIODACTYLA
FAMILY CERVIDAE

Young During the breeding season in fall, males fight fiercely for mates. Females give birth to one to three calves and care for them until due to give birth again.

Habitat Moose live in the northern parts of North America and Eurasia. They are generally called elk in Europe.

CARIBOU

Also called reindeer, the caribou is the only species of deer in which both sexes have antlers. Males are often solitary but females are more sociable and gather in herds with their young. Caribou are strong swimmers.

Close relatives
Caribou belong to the subfamily Odocoilinae, which among other species contains moose and the much smaller brocket deer and pudu.

RANGIFER TARANDUS
Caribou are preyed upon by humans, wolves, lynxes and wolverines. Their skins are used to make tents, boots and clothing and they are a source of milk and meat. They are also used as pack animals in Siberia and are associated with Santa Claus in many cultures.

Characteristics Caribou are crowned with enormous antlers, often with webs of bone between the many branches. Besides both sexes bearing antlers, caribou are unlike other Cervidae in other respects; for instance, they have a fur-covered muzzle.

Food Lichens form a major part of the caribou's diet and they will scrape away snow to get at them. Some populations range great distances between their breeding and feeding grounds.

Young In autumn, males fight over harems of females, which may number between five and forty. Female caribou have the richest milk among deer and bear highly developed, relatively large young, which can run within a few hours of birth.

Habitat Caribou inhabit the tundra of northern Europe to Asia from Scandinavia to Siberia, Alaska, Canada and Greenland.

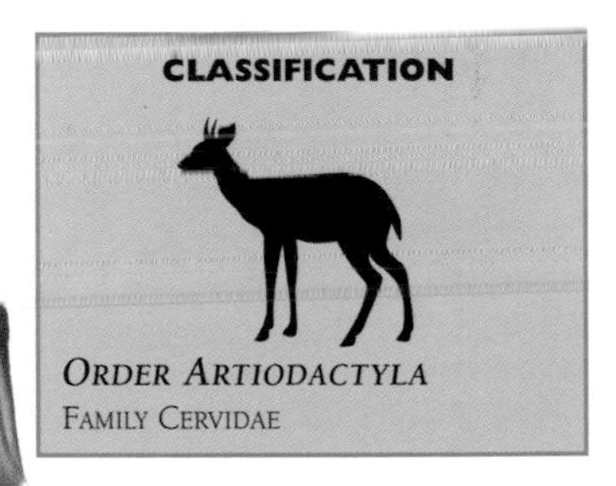

GIRAFFE

Giraffes are the largest ruminants and the tallest terrestrial mammals. They mature relatively slowly, bear few young and are potentially long-lived—up to 25 years. Females are protective mothers and kick to defend their young against hyenas, leopards and African wild dogs.

Close relatives The family contains two species: the giraffe and the okapi.

Characteristics Giraffes are fast pacers, that is, both legs on one side move more or less at the same time. Male giraffes have an average head and body length of 14 feet (4.2 m); height to the horn tips of 16 feet (5 m); and weigh 3,000 pounds (1,350 kg).

GIRAFFA CAMELOPARDALIS
Giraffes' forelegs are longer than their hindlegs. Like most mammals their necks have seven vertebrae but each one is greatly elongated. Subspecies of giraffe are distinguished by coat pattern, size and the number of horns.

Food Giraffes feed upon leaves, especially acacias. They must bend the knees and spread their forelegs widely when stooping to drink.

Young After a gestation period of 400–468 days, females give birth, usually at dawn and in a traditional calving ground, to a single, large, highly developed offspring or occasionally to twins.

"HORNED" HEAD
In giraffes, as in bovids, "horns" are formed from ossicones, but giraffes grow a covering of hair, not horn. Aging bulls progressively acquire bone material around the skull. The adult male's skull may weigh three times as much as the female's, which has few bony growths.

Habitat Giraffes are found in Africa, south of the Sahara desert.

CLASSIFICATION

ORDER ARTIODACTYLA
FAMILY GIRAFFIDAE

Okapi

"Okapi" is the name the pygmy hunters gave this animal. It was officially "discovered" in the then Belgian Congo in 1901 by the British explorer Sir Harry Johnston. Okapi live alone, each on its own home range, and meet only in the breeding season.

Close relatives The family comprises okapi and giraffe.

Characteristics The okapi has an average head and body length of about 6½ feet (2 m); height of 5–6 feet (1.5–1.8 m); and weighs about 500 pounds (230 kg). Like the giraffe, the okapi has skin-covered "horns" and lobed canine teeth. Unlike the giraffe, the okapi has scent glands on its feet. Females in heat advertise their condition by urine marking and by calling.

Food The okapi, a forest giraffe, is also a browsing, foliage feeder and, like its larger relative, gathers leaves and fruit into its mouth with an extendable, black tongue, which is so long it can also use it to clean its own eyes and eyelids.

Young Females give birth to a single young after a gestation period of 421–457 days. Calves are born in non-adult proportions with small heads, short necks and thick, long legs. They are not fully developed until four or five years of age.

Habitat Okapi are confined to the rainforests of Zaire.

CLASSIFICATION

ORDER ARTIODACTYLA

FAMILY GIRAFFIDAE

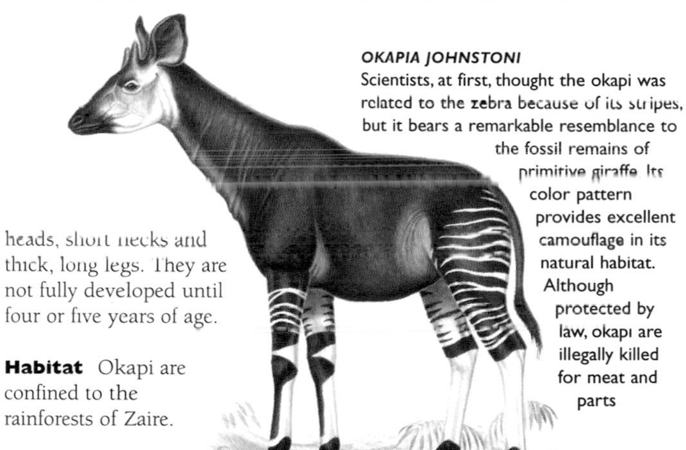

OKAPIA JOHNSTONI

Scientists, at first, thought the okapi was related to the zebra because of its stripes, but it bears a remarkable resemblance to the fossil remains of primitive giraffe. Its color pattern provides excellent camouflage in its natural habitat. Although protected by law, okapi are illegally killed for meat and parts

PRONGHORN

Pronghorns, the fastest moving mammals in North America, run with their heads held high and are specialized to move over uneven, broken terrain and able to shift their hooves quickly with each bound to avoid obstacles. Pronghorns differ from other bovids in the arrangement of the foot bones and the annual shedding of the outer horny covering of the horns.

Close relatives The pronghorn is the only species in its subfamily present in North America.

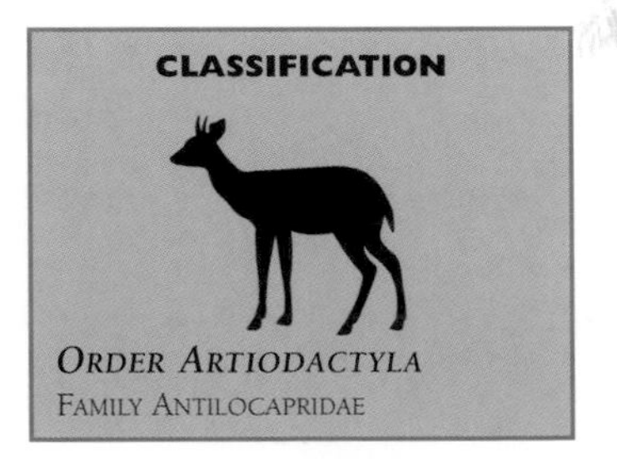

CLASSIFICATION

ORDER ARTIODACTYLA
FAMILY ANTILOCAPRIDAE

Characteristics Mature pronghorns reach an average weight of 130 pounds (60 kg), with a head and tail length of 55 inches (140 cm) and height at the shoulder of 34 inches (86 cm). The branched horn sheaths, shed immediately after mating, regrow annually. Females may also have short horns.

ANTILOCAPRA AMERICANA
Pronghorns have keen eyesight and are speedy, enduring runners. They are very light in build and deposit little fat on their bodies. The animals are strikingly marked with white underparts, throat bands and rumps on which they can erect the hairs to produce a flash of white.

Food They eat succulent forbs (herbs other than grass), shrubs particularly in winter, grasses and other plants such as cacti and domestic crops.

Young Female pronghorns produce multiple embryos that implant in the uterus. The embryos kill one another by growing long outgrowths through the bodies of rival fetuses until only two survive. Newborns hide until they are strong enough to move with the herd.

Habitat Pronghorns are found in open prairies and desert environments in central Canada, western USA and northern Mexico.

GAUR

The family Bovidae is the largest group of artiodactyls and contains about 107 species. Bovids, or hollow-horned ruminants, originated in the Old World and are currently distributed from hot deserts and tropical forests to the polar deserts of Greenland and the alpine regions of Tibet. Bovids are not indigenous to South America or Australia. The gaur is sometimes called the Indian bison.

Close relatives Gaurs belong to the tribe Bovini—the wild cattle. Yak and wild water buffalo are among their close relatives.

BOS GAURUS

It is likely that the gaur was always rather rare as it depended upon grassy glades in forests as feeding grounds. The animal has been domesticated, but the male progeny of crossbreeds with other cattle are always infertile.

Characteristics Bovids are characterized by permanent and largely hollow horns that grow from an ossicone on the forehead. Gaurs are larger than any other wild cattle and attain a shoulder height of 6 feet (1.8 m). They have white "stockings" on the legs. When alarmed, bovids snort and hold their heads high; gaurs thump the ground in unison with their forelegs as they run away.

CLASSIFICATION

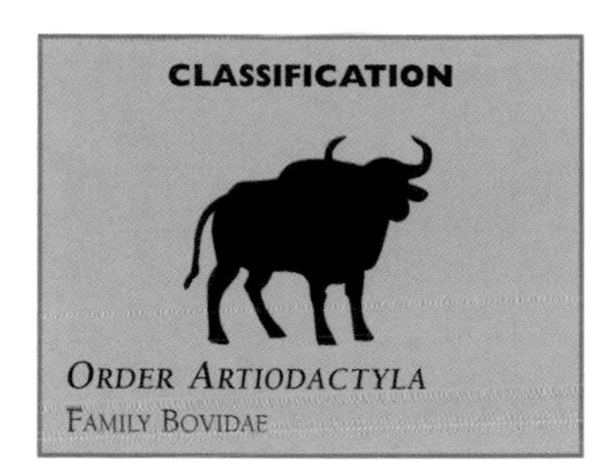

ORDER ARTIODACTYLA

FAMILY BOVIDAE

Food Gaurs graze on grasses and sometimes browse on the barks and leaves of trees. Like all ruminants, they chew the cud.

Young Females give birth to their offspring in a safe, secluded spot. They rejoin the herd a few days later.

Habitat Gaurs live in small and scattered herds in the mountain forests of India, South-East Asia and the Malay Peninsula.

Musk Ox

Musk oxen are the northernmost species of the hoofed mammals. They travel in herds of 20 to 30 animals and when attacked by Arctic wolves or dogs, will form a circle round the younger animals and present a formidable array of horns.

Close relatives Despite its size and appearance, the musk ox is more closely related to goats than it is to cattle.

Characteristics Both sexes have heavy horns that almost meet at the base. Broad hoofs prevent them sinking into soft snow. Bulls stand about 5 feet (1.5 m) at the shoulder and weigh about 880 pounds (400 kg).

Food Musk oxen eat grass and low-growing plants such as mosses and lichens and the leaves of weeping willow trees. They dig through snow to find food.

OVIBOS MOSCHATUS
Musk oxen have a somewhat ungroomed appearance on account of the long, coarse guard hairs and fine woolly underfur that protects them from the Arctic chill. In summer they shed the thick wool from beneath their shaggy coats and the Inuits use it to make fine cloth.

Young Musk oxen are so named because a musky odor is emitted from the facial glands of the bulls in the rutting season. The female produces a single calf after a gestation period of about eight months.

Habitat In the Pleistocene epoch musk oxen were circumpolar in distribution. Today, they occur in the tundra of northern Canada and Greenland; have been reintroduced successfully to Norway and Spitzbergen; and are farmed in Alaska.

CLASSIFICATION

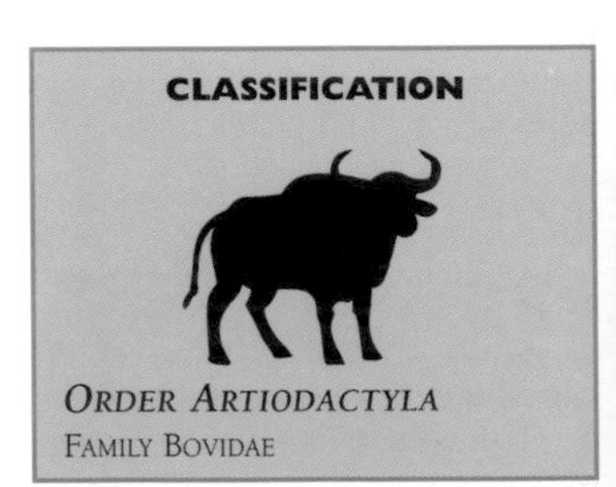

ORDER ARTIODACTYLA
FAMILY BOVIDAE

Bison

Bison are venerable beasts and have long been used by humans for meat and clothing. Hunted by land-hungry European immigrants, the American bison was almost extinct at the beginning of the twentieth century. Herds of Yellowstone and Athabasca are living wild, and other populations are preserved in a semi-wild state.

Close relatives The two types of bison in the genus are interfertile.

Characteristics The American bison may stand 6 feet (1.8 m) tall at its humped shoulder. The European bison is less heavily built with longer hindlegs. Both sexes have sharp horns.

Food American bison graze at dawn and dusk, spending the hotter hours taking dust baths or wallowing. European bison browse on leaves, ferns, twigs, bark and acorns.

Young Bison rut in the summer and births coincide with spring growth. American bison cows sometimes leave the herd to calve and lie down to give birth.

Habitat The European bison lives in semi-wild herds in Poland and the former USSR. The American bison occupies forest and grassland reserves in North America.

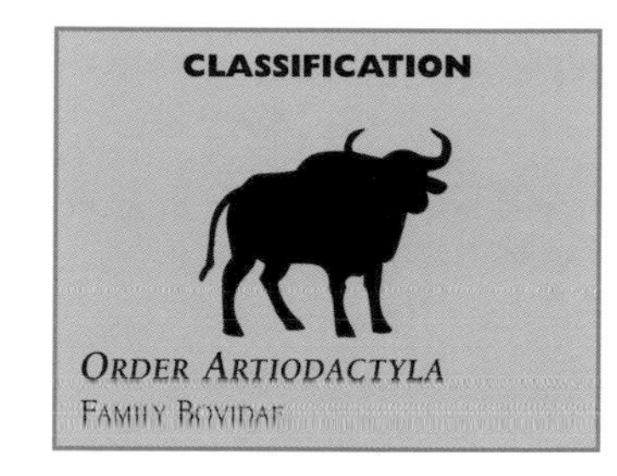

LEFT: *BISON BONASUS*
Prehistoric engravings of European bison adorn the walls of a cave at Les Cobarelles near Les Eyzies in the Dordogne region of France.

RIGHT: *BISON BISON*
The American bison's habit of stampeding is notorious. When the plains teemed with herds of these large animals, Native American hunters stampeded them over cliffs.

Four-horned Antelope

The four-horned antelope is also known as the chausinga and is, in some respects, an intermediate species between antelope and cattle. The male four-horned antelope is the only bovid that has four horns.

Close relatives The four-horned antelope is the only species in the genus *Tetracerus*.

Characteristics Four-horned antelope have a head and body length of about 40 inches (105 cm) and an overall height of about 24 inches (60 cm). The male bears two pairs of short, unringed, conical horns. The back pair are about 4 inches (10 cm) long; the front pair, which may be merely black, hairless skin, are up to $1^1/_2$ inches (4 cm) long.

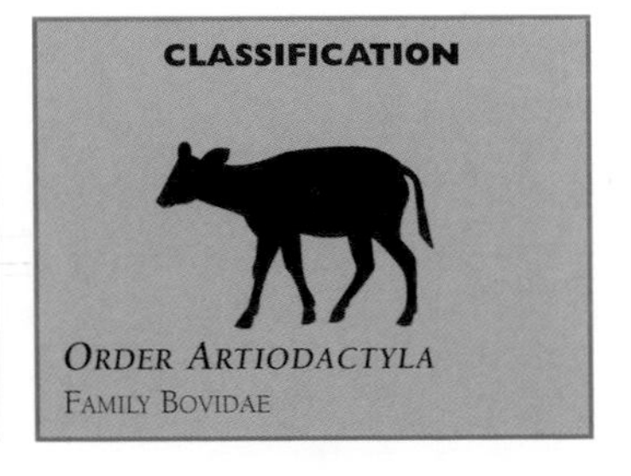

CLASSIFICATION

ORDER ARTIODACTYLA
FAMILY BOVIDAE

TETRACERUS QUADRICORNUS

This species is a relict of the stock from which the subfamily Bovinae arose. They have primitive skeletal, dental and behavioral characteristics. Normally only two adults or a female with her young are found together. When disturbed, they run for cover with a distinctive, jerky motion.

Food The four-horned antelope grazes on grasses and small plants. It drinks often.

Young Mating occurs during the rainy season. Females produce one to three young after a gestation period of about six months.

Habitat The four-horned antelope inhabits the Indian peninsula.

Bongo

The bongo is a rare medium-sized, spiral-horned antelope also called the broad-horned antelope. It is attractively marked and striped. This shy and wary animal is inactive during the day and lives in small groups or in pairs.

Close relatives The bongo shares the genus *Tragelaphus* with nyala, kudu, eland and other species.

Characteristics The bongo is the largest of the forest-dwelling antelope. Head and body length is 86–92 inches (220–235 cm); horn length is 24–39 inches (60–100 cm); tail length is 9–10 inches (23–25 cm). Females, at 463–557 pounds (210–253 kg), are lighter than males, at 529–892 pounds (240–405 kg).

Food Bongos browse at dawn and dusk on leaves, shoots, bark, rotten wood and fruit, occasionally venturing into clearings for grass.

Young One young is born after a gestation period of about 9½ months.

Habitat The bongo lives in the dense forests of central Africa from Sierra Leone to Sudan (not Nigeria) and in Kenya and Tanzania.

CLASSIFICATION

ORDER ARTIODACTYLA

FAMILY BOVIDAE

TRAGELAPHUS EURYCEROS
Lyre-shaped horns are an attribute of both the male and female bongo. When running, they align them along their slightly humped backs to prevent them catching in branches. They also use their horns to dig up roots for food.

Indian Antelope

This keen-sighted, swift antelope travels in herds of 15–20 animals. The breeding male sets up a territory and defends it and his harem against rivals. Females are especially alert to danger. When alarmed, Indian antelope begin to flee with leaps and bounds, settling into a speedy gallop over longer distances.

Close relatives The Indian antelope is the sole species in the genus *Antilope*.

Characteristics The male stands about 31 inches (80 cm) at the shoulder and has long, spirally twisted horns. Average weight is about 55–100 pounds (25–45 kg). The underparts, inner sides of the legs and an area surrounding the eyes are pure white in both sexes.

ANTILOPE CERVICAPRA
The Indian antelope is also called the blackbuck but this alternative name is misleading in some respects. Dominant males (as shown here) are black-coated only. Subordinate males have smaller horns and are yellowish fawn like the females, which have no horns at all.

CLASSIFICATION

ORDER ARTIODACTYLA
FAMILY BOVIDAE

Food The Indian antelope feeds largely on grass in the morning and evening.

Young Single young, but occasionally twins, are born after a gestation period of about six months.

Habitat This species is found on the open, grassy plains of India and Pakistan. Feral populations exist in Texas and South Australia.

SAIGA

Over time, saiga antelopes have become "short-legged" runners able to flee over level, even, hard unobstructed terrain. The horns of male saiga were once used for medicinal purposes by the Chinese, but a protection order in 1920 has allowed the numbers of this animal to increase.

SAIGA TATARICA
The saiga has a more pronounced snout than other gazelles. The downward-pointing nostrils are thought to filter dust from the air or to warm it as the animal breathes. Each nostril contains a sac, lined with mucous membranes that appear in no other mammals except whales.

Close relatives The saiga shares the tribe Saigini with the chiru or Tibetan antelope.

Characteristics The adult saiga stands about 30 inches (77 cm) at the shoulder. It runs with its head low, minimizing body lift and converting almost all its energy into forward propulsion. In winter its fawnish cinnamon coat becomes creamy white and thick and woolly.

Food Saiga feed on low-growing shrubs and grass, migrating southward in the fall to warmer, lusher pastures.

Young Females give birth in May to one to three young, which are suckled until the fall.

Habitat The saiga inhabits the colder parts of eastern to central Asia including northern Caucasus, Kazakhstan, south-west Mongolia and Sinkiang (China).

CLASSIFICATION

ORDER ARTIODACTYLA
FAMILY BOVIDAE

Klipspringer

The klipspringer is a dwarf antelope that favors rocky country, interspersed with grassy patches and clumps of bush. It moves with surefooted agility in a stilted bouncing manner on its strong legs and blunt-tipped hooves, which are the consistency of hard rubber.

Close relatives The klipspringer is the only species in the genus *Oreotragus*.

Characteristics The klipspringer stands about 22 inches (57 cm) at the shoulder. The male, and sometimes the female also, has straight spike-like horns. The tail is very short: 2½–9 inches (7–23 cm). The female is slightly heavier than the male.

Food Klipspringers feed morning and evening and on moonlit nights, principally on leaves, flowers and fruit. They also eat moss, some grasses and succulents, and drink when they can find water.

Young It is thought that klipspringers mate for life. The female may give birth twice in a year, producing a single offspring each time.

OREOTRAGUS OREOTRAGUS
As it springs from one hard surface to another, the klipspringer is cushioned from bumps by its long, thick, bristly coat. It is able to stand on its hindlegs to reach its favored foods. Klipspringers mark their territories with glandular secretions.

Habitat Klipspringers thrive in steep, rocky terrain from the Cape of Angola up the eastern half of Africa to Ethiopia and east Sudan.

CLASSIFICATION

Order Artiodactyla
Family Bovidae

Banded Duiker

While most duiker species are very common, their retiring habits ensure they are seldom seen. Duikers are considered by some authorities to be the most primitive living African antelopes. The banded duiker is a short-horned antelope that browses at the forest edge. In Afrikaans "duiker" means diver and refers to the animals' habit of diving for cover into the underbrush when frightened.

Close relatives There are 17 species of duikers in two genera.

Characteristics Duikers have the largest brains relative to body size of all antelopes. The sexes look alike, though the females are often up to 4 percent longer than the males. Banded duikers are 33–35 inches (85–91 cm) long.

Food Duikers supplement their diet of leaves, fruit, shoots, buds, seeds and bark with insects and small vertebrates and even carrion.

Young Banded duiker calves are born singly and resemble their parents in color at birth.

Habitat The banded duiker is found in Sierra Leone, Liberia and on the Ivory Coast.

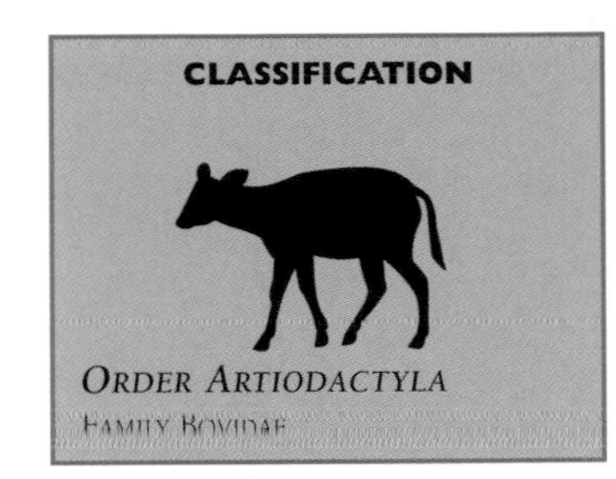

CEPHALOPHUS ZEBRA
The banded duiker is prettily marked with 12–15 black transverse stripes, which distinguish it from other species, though both the yellow-backed duiker and Jentink's duiker have attractive patterned coats. All duikers have large scent glands beneath each eye. These animals are usually seen alone or in pairs.

Blue Wildebeest

The blue or brindled wildebeest is a plains dweller that moves about in large herds. It is often known as the gnu, a name derived from the Hottentot *"t" gnu*, which describes the animal's loud, bellowing snort.

Close relatives The five subspecies of blue wildebeest share a genus with the white-tailed gnu, which is now extinct in the wild.

Characteristics Males and females look alike, but the female is smaller and lighter. Average head and body length is 76–82 inches (195–210 cm); shoulder height is 50–55 inches (128–140 cm); the tail is 17–22 inches (43–57 cm) long. Males weigh around 507 pounds (230 kg); females weigh around 352 pounds (160 kg). Wildebeest have a fast but rather ungainly gait.

CONNOCHAETES TAURINUS

Despite their size and speed, wildebeest are at great risk from lions, hyenas, cheetahs, hunting dogs and other predators. Virtually all the calves are born within a few days of one another once a year. This ensures that most of them survive.

CLASSIFICATION

ORDER ARTIODACTYLA
FAMILY BOVIDAE

Food Wildebeest eat grass almost exclusively and share feeding grounds with zebra and ostrich. They drink every two or three days.

Young The female bears a single calf after a gestation period of about eight months. It can stand within three to five minutes of birth.

Habitat Blue wildebeest are relatively abundant on the plains of central and south-eastern Africa from northern South Africa to Kenya just south of the equator.

Gemsbok

The gemsbok moves about in hierarchically organized groups of about 60 animals. Horns provide females with the means to evict competitors from scarce resources—subadult bulls may be dominated by high-ranking cows. Successful attempts have been made to domesticate the gemsbok.

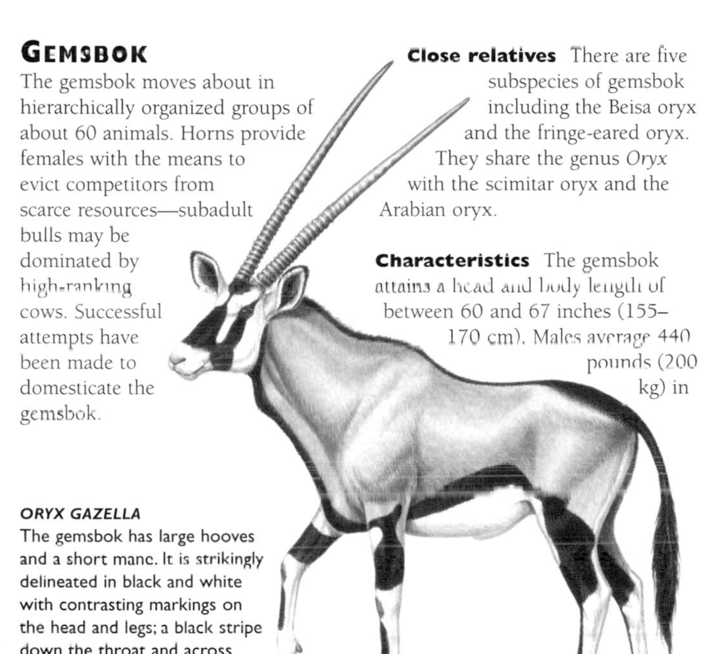

ORYX GAZELLA
The gemsbok has large hooves and a short mane. It is strikingly delineated in black and white with contrasting markings on the head and legs; a black stripe down the throat and across the flanks; and a long, black tail. Its underparts are white.

Close relatives There are five subspecies of gemsbok including the Beisa oryx and the fringe-eared oryx. They share the genus *Oryx* with the scimitar oryx and the Arabian oryx.

Characteristics The gemsbok attains a head and body length of between 60 and 67 inches (155–170 cm). Males average 440 pounds (200 kg) in weight; females are lighter at 360 pounds (160 kg). Both sexes have long, straight horns.

Food It supplements a basic diet of grass with acacia pods, wild melons and cucumbers, tubers and the bulbs of succulents.

Young A single young is concealed by the mother that visits only to nurse for the first six weeks.

Habitat The gemsbok inhabits the moister parts of the African plains.

CLASSIFICATION

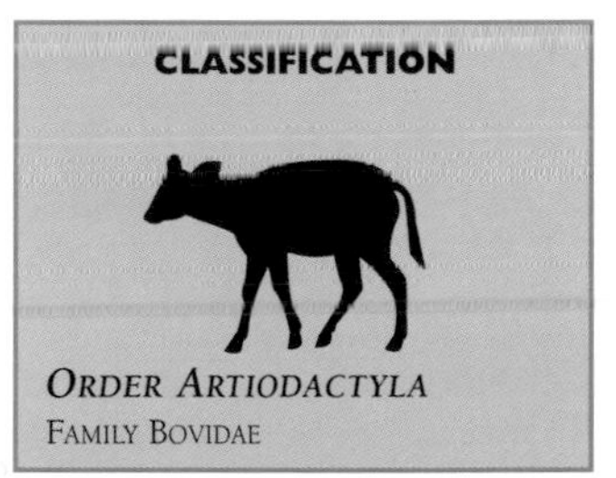

ORDER ARTIODACTYLA
FAMILY BOVIDAE

THOMSON'S GAZELLE

Gazelles, in general, are described as having a stotting or pronking gait; when alarmed or playing, they bounce along stiff-legged with all four limbs landing together. The great herds of Thomson's gazelles that occur on the African plains exhibit this behavior.

Close relatives The Thomson's gazelle is described as a small gazelle and classified in the subgenus *Gazella* with six other species.

Characteristics Thomson's gazelles, in common with other species of gazelles, are territorial in the breeding season. Males, which have much stronger horns than females, fight each other by keeping their heads low to the ground and locking horns, then pushing and twisting. They have a head and body length of 27–42 inches (70–107 cm) and a shoulder height of 15–27 inches (40–70 cm).

GAZELLA THOMSONI
Thomson's gazelle has a distinctive black stripe along its sides, which contrasts with the white underparts. There are about 15 races of this species, distinguished by color variations and horn size. Their abundant numbers are preyed upon by cheetahs, leopards, lions, hyenas, hunting dogs and humans.

Food Thomson's gazelles eat short grass with a small amount of foliage. Grazing supplies most of their moisture needs; they drink only when the vegetation is dry.

Young Females calve at any time and may produce two offspring in a year. The gestation period for all gazelles is $5^{1}/_{2}$ months.

Habitat Thomson's gazelle is distributed through Tanzania and Kenya and southern Sudan.

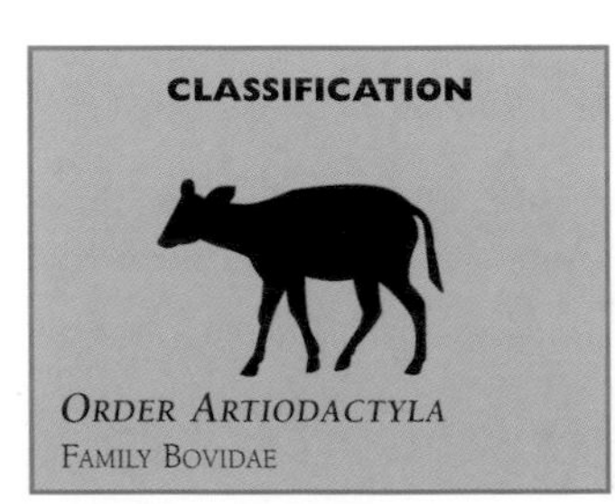

GERENUK

The gerenuk's name comes from the Somali for "giraffe-necked." It has an unusually long neck and long, slim legs. Rearing on the hindlegs is a characteristic of many gazelles. The gerenuk is able to maintain this stance for some time and even to walk on its hindlegs to a certain extent, a useful skill when foraging for food.

Close relatives Relatives are the gazelles and impala.

Characteristics The gerenuk has a less complicated digestive tract and weaker molar teeth, with a smaller grinding surface, than some other species of gazelles. The females lack horns, which makes their long, pointed ears look all the more distinctive. Gerenuks are 31–40 inches high (80–105 cm) at the shoulders.

Food The gerenuk is largely a solitary browser. It feeds selectively on leaves, shoots and fruit and obtains enough water from its food sources to make drinking almost unnecessary.

Young Females usually give birth to single offspring in the rainy season.

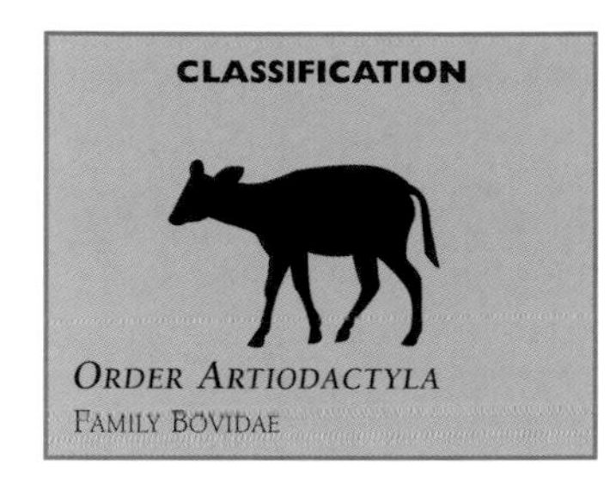

Habitat The gerenuk inhabits dry environments, from desert to bush savanna, in the Horn of Africa south to Tanzania.

LITOCRANIUS WALLERI
Gerenuk browse morning and evening, often propped against a tree trunk on their hindlegs, using a foreleg to pull down branches and stretching their long neck. They stand still in the shade when the sun is high, ever alert for danger. Their main enemies are the cheetah, leopard, lion, hyena and hunting dog.

WILD GOAT

The goat antelopes and their descendants in the subfamily Caprinae developed in the ice ages into a large diversity of species.

Close relatives
There are many races in the genus *Capra* to which the wild goat belongs.

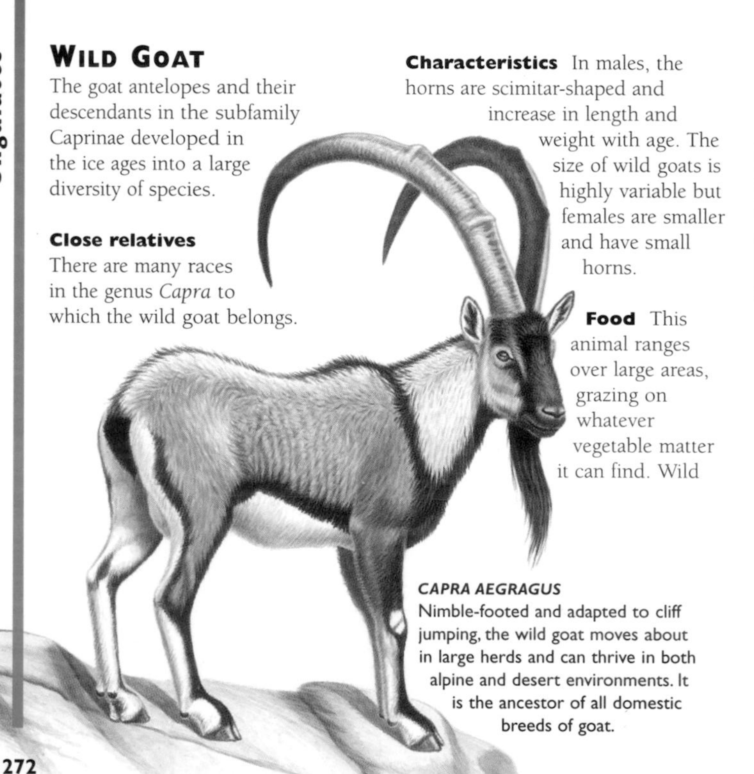

Characteristics In males, the horns are scimitar-shaped and increase in length and weight with age. The size of wild goats is highly variable but females are smaller and have small horns.

Food This animal ranges over large areas, grazing on whatever vegetable matter it can find. Wild goats become ruthlessly territorial when food is scarce.

CAPRA AEGRAGUS
Nimble-footed and adapted to cliff jumping, the wild goat moves about in large herds and can thrive in both alpine and desert environments. It is the ancestor of all domestic breeds of goat.

CLASSIFICATION

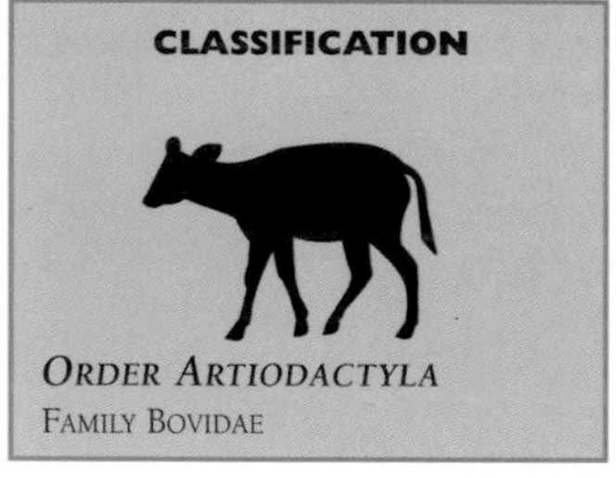

ORDER ARTIODACTYLA
FAMILY BOVIDAE

Young Twins are more common than in other related species, which may account for its success as a feral and domesticated animal.

Habitat The wild goat is widely distributed through the Greek Islands, Turkey, Iran, south-west Afghanistan, Oman, Caucasus, Turkmenia, Pakistan and India. Feral and captive domestic goats are found worldwide.

Bighorn Sheep

The bighorn sheep lives in herds in inaccessible mountainous areas, especially on steep cliffs. In summer, the rams live apart from the ewes; they use their horns like sledge-hammers when they fight for the females' favors.

Close relatives The Canadian bighorn and Dall sheep are closely related and share the genus with the urial, argalis, mouflon and snow sheep.

Characteristics The genus *Ovis*, to which bighorn sheep belong, are characterized by preorbital, foot and groin glands. There is a great range in body weight, which varies between 125 and 310 pounds (60–140 kg), depending on locality.

Food It feeds mainly on grasses, but will eat shrubs and even cactii.

Young Bighorn sheep breed in November and December. Single or twin lambs are born after a gestation period of six months. The female has two teats and tends her young carefully.

Habitat Bighorn sheep inhabit south-west Canada to western USA and north Mexico where they are found in a wide variety of habitats ranging from cold alpine areas to dry deserts.

Classification

Order Artiodactyla
Family Bovidae

OVIS CANADENSIS
The horns of male bighorn sheep are large, forward-winding and impressive. Both sexes have large rump patches and short tails. The desert bighorns are susceptible to the diseases of domesticated livestock and these, together with loss of habitat, have severely reduced their numbers.

Kinds of Mammals

Aardvark and Pangolins

The aardvark and pangolins are among the strangest of animals. The aardvark is the only mammal classified in an order of its own—Tubulidentata. Believed to have evolved from an early hoofed mammal, the aardvark exhibits primitive anatomical and molecular characters and is thought of as a "living fossil." At one time classified with the anteaters, sloths and armadillos, the pangolins, too, have been placed in their own distinctive order—Pholidota. Distinguishable from all other Old World mammals by their unique covering of horny body scales, pangolins are truly extraordinary-looking animals that, at first glance, look more reptilian than mammalian.

AARDVARK

These nocturnal, solitary and elusive creatures are rarely observed in the wild. The order name comes from the primitive, oval teeth that are flat on top and columnar, with pulp cavities that appear as tubes on the chewing surfaces.

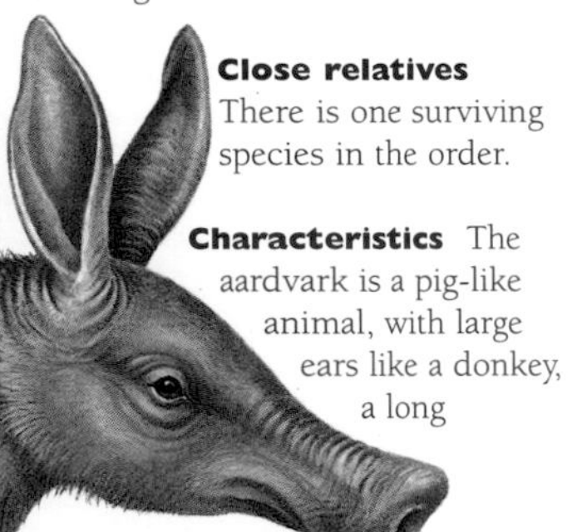

Close relatives There is one surviving species in the order.

Characteristics The aardvark is a pig-like animal, with large ears like a donkey, a long muscular tail, and a nasal cavity with nine to 10 thin bones—more than any other mammal. Hearing and smell are acute, but the eyes are reduced. The skin is thick and sparsely covered in bristly hair. With its massive skeleton, thick skin and sharp, shovel-shaped claws, the aardvark is well armed. Aardvarks dig separate burrows for food, temporary shelter and permanent residence.

ORYCTEROPUS AFER
An aardvark can excavate a termite mound with powerful, strongly clawed forefeet in a few minutes, where a human would need to use an ax. The nostrils can be closed by a muscular contraction of hair to prevent insects entering.

Food The aardvark feeds on ants, termites, locusts and grasshoppers.

Young After seven months' gestation the young are born with well-developed claws and eyes open. They join their mothers on nocturnal foraging trips from two weeks of age and remain with them until the next mating season.

Habitat Aardvarks are found wherever termites are found: open canopy forests, bush veldts and savannas.

Tubulidentata distribution

Aardvarks are known as fossils from the Pliocene in Europe and Asia, and in Africa only from the Miocene to recent times. Today, they are confined to Africa. They range from south of the Sahara, through the equatorial countries to South Africa.

CLASSIFICATION

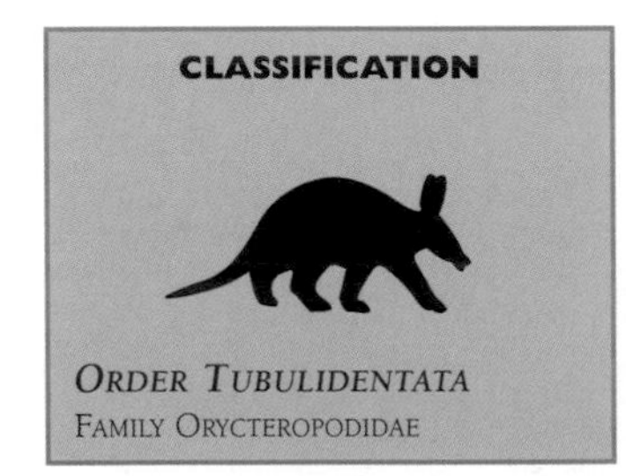

ORDER TUBULIDENTATA
FAMILY ORYCTEROPODIDAE

PANGOLINS

Like aardvarks, pangolins are solitary, shy, noctural insectivores, with an acute sense of smell. Unlike them, pangolins have no teeth, but a 27 inch (68 cm) tongue. Unlike any other Old World mammal, they are covered in body scales.

Close relatives There are three Asian and four African pangolin.

Characteristics All pangolins have short, powerful, digging limbs. The scaly tail is mobile and very sensitive at the tip. It can be hooked like a finger over a solid support, or lashed at an adversary, using the razor-sharp scales to devastating effect. This action may be supplemented by spraying a foul smelling liquid from the anal glands.

Food Pangolins feed exclusively on termites and ants.

Young A single young grasps the base of the mother's tail and when threatened, she wraps the tail around herself, enclosing the baby.

Habitat Pangolins conceal themselves in underground burrows or tree hollows during the day. Their tropical rainforest habitat is essential to the survival of pangolins.

Pholidota distribution

Pangolins occur in much of South-East Asia, and in tropical and subtropical Africa.

MANIS SP.
The three Asian pangolins possess external ears and have a scale-clad tail and hairs at the base of the body scales. The Chinese have traditionally attributed medicinal values to the scales and the animals have been relentlessly hunted.

Kinds of Mammals

Rodents

More than one-third of the world's mammals belong to the order Rodentia. They have 443 genera in 29 families. Their great success, despite having many predators and few means to defend themselves, is due to their adaptability and high reproduction rates. The 2,000 or so species are remarkably similar in appearance, differing mainly only in adaptations to feeding habits (in teeth and digestive systems) and habitat (in limbs and tails). Their relationships with humans are often close and frequently deleterious: historically they have spread fatal diseases on an enormous scale, and they can consume large quantities of crops and stored foodstuffs.

THE WORLD OF RODENTS

Rodents have remained relatively unspecialized throughout their evolution. The combination of this evolutionary flexibility with small size and high reproduction rates has contributed to their success. Small structural and functional changes are enough to produce the diverse array of modern species.

A simple plan Rodents have a remarkably uniform mouse-like body plan that has been modified mainly in the teeth, digestive system, limbs and tail. Most rodents weigh less than 5 ounces (150 g)—though there are exceptions—enabling the exploitation of a wide range of microhabitats. The short gestation periods, large litters and frequent breeding patterns of rodents permit survival under adverse conditions and rapid exploitation of the species under favorable ones.

NEW WORLD CAVY-LIKE RODENTS
The Caviomorpha are the most diverse group of rodents. This is probably because they evolved early—57 to 37 mya—in the Northern Hemisphere and crossed into South America 37 to 24 mya. The capybara is a recent species, evolving only around 5 mya.

OLD WORLD MOUSE-LIKE RODENTS
The black rat appears to have originated in western Asia. It probably evolved from ancestral stock arriving there between 24 to 5 mya. It then diversified and spread to establish secondary evolutionary centers in Africa and Australia.

NEW WORLD MOUSE-LIKE RODENTS
The lemming arose from hamster-like stock around 37 mya. The establishment of a land bridge between North and South America gave these ancestors their greatest opportunity to radiate.

Suborders Three major groups of rodents are separated on the basis of their jaw muscles. Their functions vary depending on the position of the muscle branches. The parts of the skull associated with these different muscle functions are quite distinct for each of the groups. The three rodent groups are known as the Sciuromorpha, or squirrel-like rodents; the Caviomorpha, or cavy-like rodents; and the Myomorpha, or mouse-like rodents.

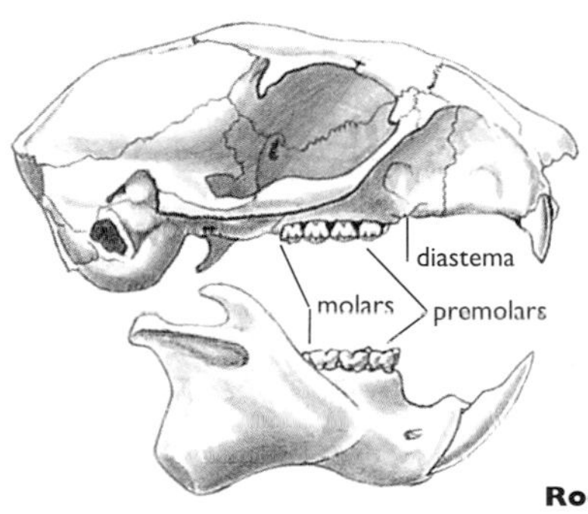

IDENTIFYING FEATURES OF RODENTS

The skull of a rodent can be identified by a pair of continuously growing, chisel-edged incisors in the upper and lower jaws, and a long gap between these and the grinding teeth. The diastema permits the lips to be brought together to exclude unwanted particles of gnawed material. There are no canines.

OLD WORLD CAVY-LIKE RODENTS

The crested porcupine is one of three species of Old World porcupines. It is a member of the *Hystrix* genus, which have shorter tails and longer quills than their relatives. They occur from Java and Borneo, through southern Asia to southern Europe and through most of Africa.

Rodent distribution

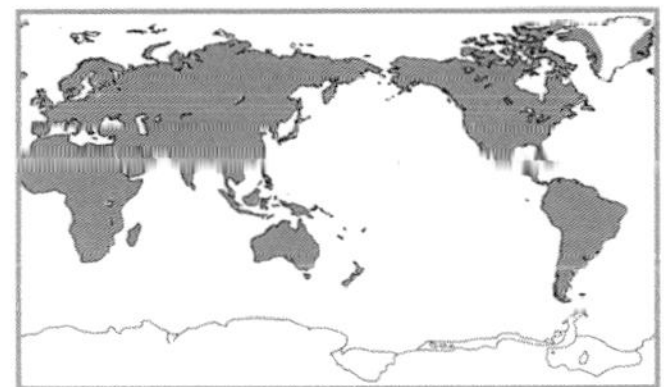

Rodents have a worldwide distribution. They are found from the high Arctic tundra to tropical deserts, forests and mountains. They have also reached some of the most isolated oceanic islands.

Muskrat

The Muridae did not appear until Pliocene times, which makes them relatively young. With more than 1,000 species, this family is still maximizing its genetic diversity. The muskrat spends much of its life in water and its hindfeet are partially webbed and fringed with stiff bristles. It is also known as a musquash, an Algonquin name thought to mean "big rat."

Close relatives The vast family contains rats, mice, voles and lemmings.

Characteristics The muskrat's compact head and body is about 12 inches (30 cm) long. The tail is long and scaly and flattened from side to side. The coat is dense, soft underfur thickly overlaid with long, stiff, glossy guard hairs.

CLASSIFICATION

Suborder Myomorpha
Family Muridae

Food The muskrat feeds mainly on the stems and roots of water plants but occasionally supplements its diet with freshwater mussels, crayfish, salamanders and fish.

ONDATRA ZIBETHICUS
The muskrat digs burrows in the banks of waterways and builds mounds in water constructed from reeds, rushes, small branches and twigs. This animal is named for the musky-smelling secretion that comes from musk sacs in the anal region. Its fur is used in the fur industry; its flesh is eaten as "marsh rabbit."

Young The muskrat breeds all year round in the southern part of its range, but only in summer elsewhere. Litters vary in size from one to eleven. The gestation period is three to four weeks.

Habitat It is found over most of North America except in treeless tundra and has been introduced into parts of Europe.

GERBIL

Gerbillinae are a large subfamily of murids containing some 70 species and distributed across central and western Asia and Africa. The Indian gerbil lives in communities in deep burrow systems with many entrances. It often blocks burrow entrances loosely with soil to deter snakes and mongooses.

Close relatives These are the typical rats and mice.

Characteristics Water loss from the lungs is problematic for desert-dwelling animals. Gerbils have specialized nasal bones, which condense water vapor from the air before it is expired, allowing it to be re-absorbed into the system. Indian gerbils' tails are slightly longer than their bodies and tufted at the end.

Food The diet consists of green vegetation, bulbs, roots, insects, eggs and nestlings.

Young Indian gerbils breed all year round. They have litters of up to eight young.

CLASSIFICATION

SUBORDER SCIUROMORPHA
FAMILY MURIDAE

Habitat Indian gerbils inhabit desert and semi-desert regions of western and southern Asia.

TATERA INDICA
These well-adapted desert-dwellers have efficient kidneys, which concentrate urine and thus conserve moisture; long hindlegs to raise bodies above the hot sand; dense pads of fur to insulate the soles of their feet; and white underbellies to reflect radiated heat.

Black Giant Squirrel

Arboreal squirrels usually have a long counterbalancing tail to facilitate movement through the trees. They jump with the limbs outstretched, the body flattened and the tail slightly curved. They are agile, live in tree hollows or nests (dreys) built of leaves and twigs and are usually active throughout the year.

Close relatives The family Sciuridae is enormously diverse, containing 267 species in 49 genera.

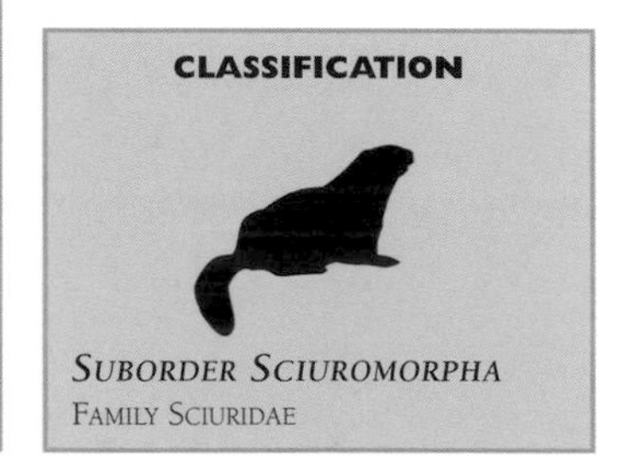

CLASSIFICATION

Suborder Sciuromorpha
Family Sciuridae

Ratufa bicolor

All squirrels hold food in their front paws. The black giant squirrel can leap 20 feet (6 m) or more through trees. Its tail is a balance for running and climbing, a rudder for jumping, a flag for communicating social signals, and a wrap when the animal sleeps.

Within the family there are four species of giant squirrels.

Characteristics Tree squirrels have five toes on the hindfeet, excellent eyesight and wide vision; they can appreciate distance in three dimensions. The black giant squirrel's tail is slightly longer than its head and body length of 11–18 inches (28–45 cm). It is handsomely marked and weighs up to $6\frac{1}{2}$ pounds (3 kg).

Food The black giant squirrel supplements its main diet of fruit, nuts and bark with the occasional small invertebrate.

Young A pair of black giant squirrels makes a huge nest in the breeding season. This is shelter for one or possibly two offspring, which are born after a gestation period of about four weeks.

Habitat The black giant squirrel is found in dense forest from Nepal to Indonesia.

European Red Squirrel

The European red squirrel's arched leaping movement on the ground is typical of arboreal squirrel species. This species was plentiful in Britain until the arrival of the gray squirrel; it is now present in greatly reduced numbers. Both red and gray squirrels are popular animal characters in folk stories and children's literature.

NUT CRACKING TECHNIQUE

Gripping a nut with both hands and its upper incisors, the squirrel gnaws through the shell with its lower incisors, then uses them to lever the shell apart. Young squirrels soon learn to distinguish between sound and wormy nuts.

Close relatives There are two related species in Asia, but another 24 in North and South America.

Characteristics The red squirrel has a head and body length of 8–9 inches (20–23 cm), a tail of 6–8 inches (15–20 cm) and conspicuous ear tufts. Like many other species, the red squirrel has a well-developed sense of touch with touch-sensitive whiskers, called vibrissae, on the head, feet and the outside of legs.

Food Conifer cones, acorns and other nuts are its main food but it also eats fungi and fruit in summer.

Young In a good year, females may produce two litters of about three young. A drey is used for raising offspring and as winter quarters.

Habitat The European red squirrel is distributed in coniferous forest through Europe, east to China, Korea and Hokkaido in Japan.

SCIURUS VULGARIS
European red squirrels damage forestry plantations by eating young conifer shoots, but their failure to recover cached acorns disperses oak trees. Adults can smell and relocate pine cones buried 12 inches (30 cm) deep.

CLASSIFICATION

SUBORDER SCIUROMORPHA
FAMILY SCIURIDAE

Gray Squirrel

Introduced into Britain in 1876, this prolific breeder is ousting the native European red squirrel from the woodlands of southern England. The hairs from the tails of gray squirrels are used in artists' brushes.

Close relatives The gray squirrel's close relatives are other species of tree squirrels, flying squirrels and ground squirrels.

Characteristics Sharp-toed tree squirrels are able to hold themselves motionless and head-down on tree trunks while waiting to make their next move. The gray squirrel is 9–12 inches (23–30 cm) long with a tail length of 8–9 inches (20–23 cm).

Food The gray squirrel eats seeds, tree-sap and nuts; an adult can consume 3 ounces (80 g) of nuts a day. It also eats eggs, young birds and insects.

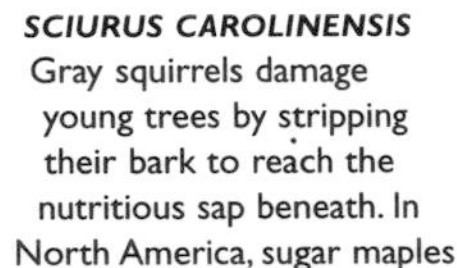

SCIURUS CAROLINENSIS
Gray squirrels damage young trees by stripping their bark to reach the nutritious sap beneath. In North America, sugar maples are most at risk; in Britain, they attack mostly sycamores, beech and oak. Gray squirrels are preyed upon by owls, foxes and bobcats, which keep the numbers down.

CLASSIFICATION

Suborder Sciuromorpha
Family Sciuridae

Young Two litters are produced each year in early spring and summer. Though seven young may be born, usually only three or four survive. The male takes no part in rearing the young and if the mother senses danger, she will move her young to another nest, carrying them by the scruff of the neck.

Habitat Its wild range is hardwood forest in south-east Canada and eastern USA. The species has been introduced in Britain, Ireland and South Africa.

Tassel-eared Squirrel

This arboreal species is so named for its distinctive ear tufts. It has a gray to reddish back and the tail fur is darker on the top than beneath.

Close relatives This squirrel shares the genus *Sciurus* with about 27 other species.

Characteristics The tassel-eared squirrel is shaped like many other tree squirrels with a long cylindrical body and a long, bushy tail. In common with many other species, tassel-eared squirrels do not need copious supplies of drinking water and can survive on moisture from their food except in hot summers.

Food Nuts and seeds form the principal food and a highly nutritious staple diet.

Young Like all squirrel species, the gestation period is short and the young are born in nests or dreys.

Habitat The tassel eared squirrel inhabits yellow pine forests above 7,000 feet (2,100 m) in the southern Rocky Mountains and Grand Canyon area of south-western USA.

CLASSIFICATION

SUBORDER SCIUROMORPHA
FAMILY SCIURIDAE

SCIURUS ABERTI
Sciurids have short forelimbs, with a small thumb and four toes on the front feet, and longer hindlimbs with five toes on the hindfeet. Their sharp claws are not only useful for dealing with their food—they also provide excellent traction allowing the animal to run vertically as well as horizontally.

Southern Flying Squirrel

Flying or gliding squirrels are nocturnal and are less agile than the diurnal arboreal species. On landing, they brake by flexing the body and tail upward.

Close relatives The southern or least flying squirrel is most closely related to the northern flying squirrel and other species of flying squirrel.

Characteristics Flying squirrels control direction in the air by the position of the legs and tail and the stiffness of the flight membrane. They launch into a glide when martens appear, but are less successful at avoiding flying predators such as owls.

GLAUCOMYS VOLANS
Flying squirrels do not actually fly. Their gliding movement is facilitated by the patagium—a membrane down each side of the body connecting their forelegs and hindlegs. They use this as a parachute and the tail as a rudder.

Food This species forages in the tree-tops for nuts, bark, lichens, fungi, fruit and berries and hoards food in its nest. It stores nuts and dried berries in hollow trees for use during the winter months.

Young Like all infant squirrels, baby southern squirrels are born naked, toothless and with their eyes closed, but with a well-developed gliding membrane. They suckle for about 60–70 days until they are physically developed enough to attempt gliding.

Habitat The southern flying squirrel inhabits deciduous and mixed woodlands and forests from the eastern USA and south-eastern Canada to Honduras.

Columbian Ground Squirrel

Most ground squirrels live in small social groups sharing a burrow system where the young are raised, food is stored and refuge is sought from predators. Co-operative alarms are raised when danger threatens. Many temperate species of ground squirrels hibernate.

Close relatives There are 36 species of ground squirrels in the genus *Spermophilus*.

Characteristics Ground squirrels have strong forelimbs and long claws suited to digging. Their tails are much shorter than in other squirrel species. Ground squirrels have cheek pouches for carrying food, which they store in their burrows. Columbian ground squirrels are active only for four months of the year.

Food Columbian ground squirrels eat herbs and grasses close to their burrows.

Young Young are born late May to mid-June and emerge from the communal burrows 21 to 29 days later. They grow rapidly and are almost full adult size when they hibernate in September.

Habitat It inhabits alpine meadows and grassy riverbanks in the northern USA and southern Canadian Rockies.

SPERMOPHILUS COLUMBIANUS
Northern Hemisphere ground squirrels hibernate in winter. They prepare for hibernation by becoming quite fat and by storing food in their dens, which are generally used only when the animals awaken in the spring.

Least Chipmunk

Chipmunks are small, terrestrial rodents that live primarily on the ground. They are also able to climb trees and can swim well. The least chipmunk is the most common animal seen around campgrounds in western North America.

Close relatives Chipmunks belong to the squirrel family and comprise two genera.

Characteristics
Chipmunks have capacious cheek pouches for carrying food. They may sleep for extended periods during the winter, but do not truly hibernate. They call with a shrill chirr.

Food
Chipmunks are early morning foragers. Their preferred diet is nuts, berries and a wide assortment of seeds.

TAMIAS MINIMUS

The least chipmunk, like other chipmunk species, can be tamed and kept as a pet. Its natural habitat is semi-open land where it often makes a nuisance of itself by uprooting newly planted corn seed in spring and raiding granaries in fall.

CLASSIFICATION

SUBORDER SCIUROMORPHA
FAMILY SCIURIDAE

Young Female chipmunks bear two to eight young in spring or summer after a gestation period of about a month. A second smaller litter may be produced in areas where the summers are long.

Habitat Least chipmunks are found across Canada from Yukon to Ontario and in the Rockies and Great Basin of the western USA. They inhabit diverse environments from alpine tundra, coniferous forests and aspen groves, down to the sagebrush desert.

Black-tailed Prairie Dog

Prairie dogs live in social units, called coteries, consisting of an adult male, several adult females and their young. Resources are shared within the territory and no one animal is dominant within the coterie. Several coteries form "towns," which may cover areas of up to 160 acres (65 ha). New territories are set up when adults move out of an established coterie. The burrows are often surrounded by crater-shaped mounds of earth to help prevent flooding.

Close relatives There are five species of prairie dogs.

Characteristics The black-tailed prairie dog has short legs. It is 12–17 inches (30–43 cm) long including a 1–5-inch (3–12-cm) tail. Animals within coteries frequently touch and groom each other.

Food Prairie dogs emerge from their burrows in the daytime to graze on grass and other vegetation. Their liking for herbs assists the growth of grama grass.

Young Litters of up to 10 young are born in March, April or May. They are weaned at about seven weeks of age.

Habitat The black-tailed prairie dog occurs over the plains and plateaus of America, and ranges from southern Saskatchewan south to northern Mexico.

CYNOMYS LUDOVICIANUS

The black-tailed prairie dog looks a little like a stocky terrier and utters a sharp dog-like bark when it senses danger. Its natural enemies are eagles, foxes and coyotes. Prairie dogs bite off all tall growth in the vicinity of their burrows to open up a field of view and keep lookout from the mounds at burrow entrances. Prairie dogs may damage crops and compete with domesticated livestock for grass.

CLASSIFICATION

SUBORDER SCIUROMORPHA

FAMILY SCIURIDAE

HOARY MARMOT

Marmots are terrestrial and have a similar social structure to ground squirrels. The hoary marmot is also known as the whistler because of its piercing alarm call. It is hunted locally for its fur and for food and its hide is made into shoelaces.

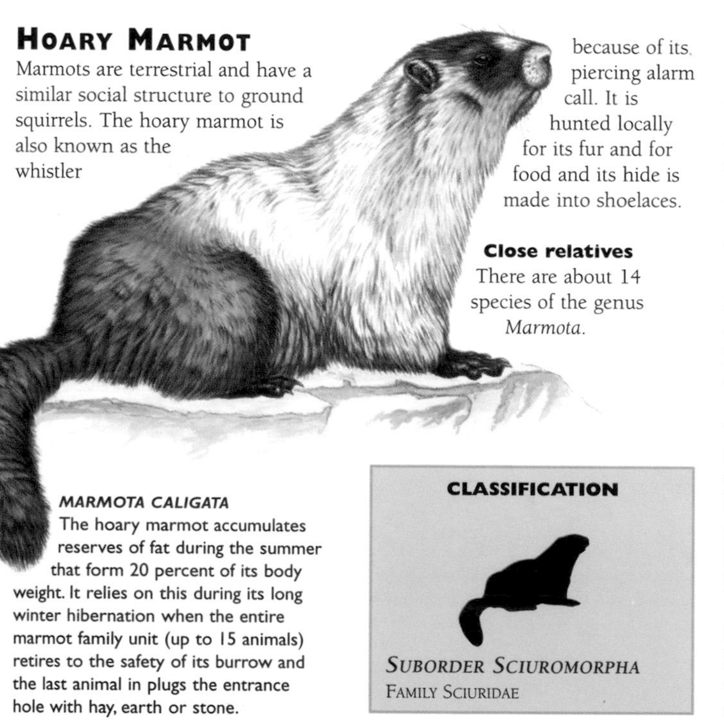

Close relatives There are about 14 species of the genus *Marmota*.

MARMOTA CALIGATA
The hoary marmot accumulates reserves of fat during the summer that form 20 percent of its body weight. It relies on this during its long winter hibernation when the entire marmot family unit (up to 15 animals) retires to the safety of its burrow and the last animal in plugs the entrance hole with hay, earth or stone.

CLASSIFICATION

SUBORDER SCIUROMORPHA
FAMILY SCIURIDAE

Characteristics Marmots are heavyset rodents with coarse fur, small ears, short tails and strong feet and claws adapted for digging. The hoary marmot is distinguished by mixed black and white fur on its head and shoulders—giving rise to the name hoary—and black "boots."

Food Marmots are diurnal feeders and live almost entirely on green plants, sometimes doing considerable damage to crops.

Young Most squirrels are sexually mature and able to breed within a year; marmots are not fully grown until they are two years old. They mate soon after they wake from hibernation and litters of four or five are born.

Habitat The hoary marmot lives high in the mountains of north-western North America, northern Canada and Alaska.

Plains Pocket Gopher

Pocket gophers are so named for their deep, fur-lined cheek pouches in which they carry food. They live alone, except during the breeding season, and spend much of their lives underground, digging complex burrow systems over wide areas. They do not hibernate, but hoard food in order to survive the winter.

Close relatives There are some 30 to 40 species of pocket gophers distributed through North and Central America.

Characteristics The plains pocket gopher has small eyes and ears and chisel-like incisor teeth. Its strong, broad paws have well-developed front claws for digging. It is 7–10 inches (18–25 cm) long with a tail 4–5 inches (10-12 cm) long.

Food Pocket gophers eat surface vegetation and underground roots and tubers and sometimes become serious agricultural pests. Food storage areas are usually sealed from the main tunnel system.

Young Gophers breed in spring or early summer. Litters contain two to three young, which are weaned at 10 days, but remain in their mother's burrow until they are about two months old. Plains pocket gophers are sexually mature at three months.

Habitat The plains pocket gopher is found in sandy soil in sparsely wooded areas in central USA from the Canadian border to Mexico.

GEOMYS BURSARIUS
Despite its crop-damaging potential, the pocket gopher's burrowing habit has positive outcomes. The burrows aerate and add organic matter to the soil and also collect runoff water from melting snow. In the long term, therefore, gopher activity improves the productivity of pastureland.

Spiny Rat

The family Echimyidae includes some soft-furred species but most have hair that is stiff and spiny. Although not at all closely related to porcupines, the spiny rats suggest how quills may have evolved from ordinary hairs.

Close relatives Other species in the genus *Proechimys* are related; all look very alike.

Characteristics The head and body length of spiny rats ranges from 6 to 16 inches (15–40 cm) excluding the tail, which varies in length and the amount of hair covering. Most spiny rats have pointed noses. They may be reddish brown, brown, blackish, black and white.

***PROECHIMYS* SP.**
Spiny rats live in forests and clearings and shelter in burrows, among rocks or tree roots and in tree hollows. They are active in the evening and during the night. Some species are terrestrial; others spend most of their lives in trees. They are hunted locally by humans for food.

Food Spiny rats are herbivores. Their diet consists mainly of sugarcane, grass, nuts and fruit.

Young Little is known about the breeding behavior of spiny rats.

Habitat Spiny rats are found in Central and South America, frequently near water.

CLASSIFICATION

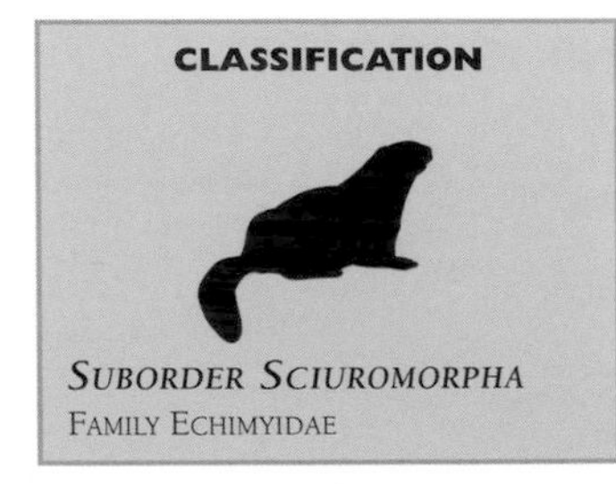

SUBORDER SCIUROMORPHA
FAMILY ECHIMYIDAE

AMERICAN BEAVER

Beavers are great builders, using their large, ever-growing incisor teeth to fell trees for food and for building dams across streams to impound water and create ponds. They build conical lodges in the ponds with access to the living chamber through an underwater tunnel. The liquid castoreum from the musk glands of both sexes is an ingredient in some perfumes.

Close relatives The American beaver is closely related to the European beaver, the only other species in the genus *Castor*.

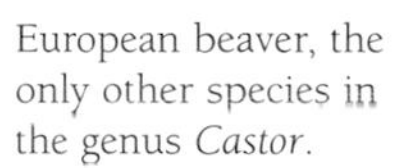

CASTOR CANADENSIS
American beavers live in hierarchical family units, a social system unique among rodents. Each family, which consists of an adult pair and the offspring from several previous years, occupies a discrete, individual territory.

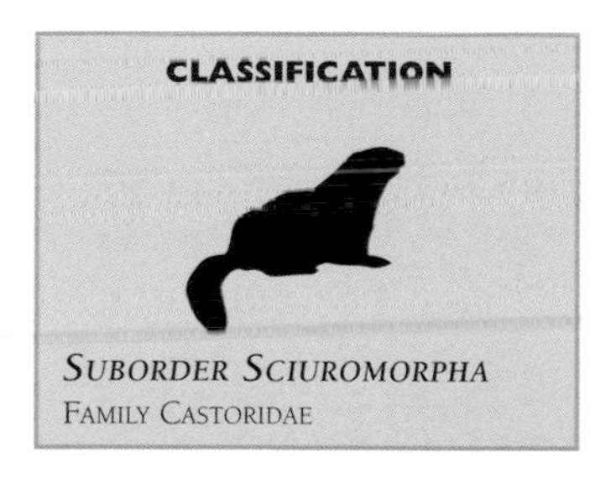

CLASSIFICATION

SUBORDER SCIUROMORPHA
FAMILY CASTORIDAE

Characteristics The streamlined body, flattened, scaly tail and webbed hindfeet make the beaver ideally suited to semi-aquatic life. Their stomachs are specially adapted for digesting woody plant material. Beavers may grow to 4 feet (1.3 m) long from nose to tail tip and may weigh more than 60 pounds (27 kg).

Food Beavers are herbivores, preferring the tender bark and buds of trees, ferns, grasses and other non-woody plant material in spring and summer. They store food in preparation for the winter.

Young American beavers have one litter a year of up to eight kits. Parents are attentive carers.

Habitat The American beaver is distributed from Alaska to Mexico. It has been introduced to Tierra del Fuego and is spreading in Patagonia.

Crested Porcupine

It has been suggested that the first Caviomorphs evolved from a Northern Hemisphere ancestor during the Eocene epoch. The crested or African porcupine, also called the quill pig, is the largest terrestrial rodent. It is mainly nocturnal, usually solitary and digs a burrow for rest and refuge.

Close relatives The 11 species of Old World porcupines are divided into two subfamilies and four genera.

Characteristics This animal attains a head and body length of about 32 inches (80 cm) and a weight of about 60 pounds (27 kg). The tail is relatively short and surrounded by cylindrical, stout, sharp quills. The head is crowned with long, coarse hair that can be erected.

HYSTRIX CRISTATA
When threatened, the crested or African porcupine grunts, stamps its hindfeet, raises its back quills and rattles those on its tail. It then attacks by running sideways or backward into the enemy. On contact, the quills detach easily; embedded quills often cause septic wounds.

CLASSIFICATION

Suborder Caviomorpha
Family Hystricidae

Food It eats a vegetarian diet, taking roots, bulbs, fruit and berries from a variety of plants, some of which are poisonous to cattle.

Young The young are born in a special chamber of the excavated burrow system. Females sometimes breed twice a year, producing one to four young. Female mammary glands are placed on the side of the body.

Habitat This species is widely distributed throughout Africa, Italy and the Balkans.

Prehensile-tailed Porcupine

After North and South America reconnected in the Pliocene epoch only one Caviomorph, a porcupine, moved north and successfully established itself. Normally, the spines of prehensile-tailed porcupines lie smoothly along the back but are erected by muscles in the skin when the animal senses danger.

Close relatives Six of the nine species of New World porcupines have prehensile tails.

CLASSIFICATION

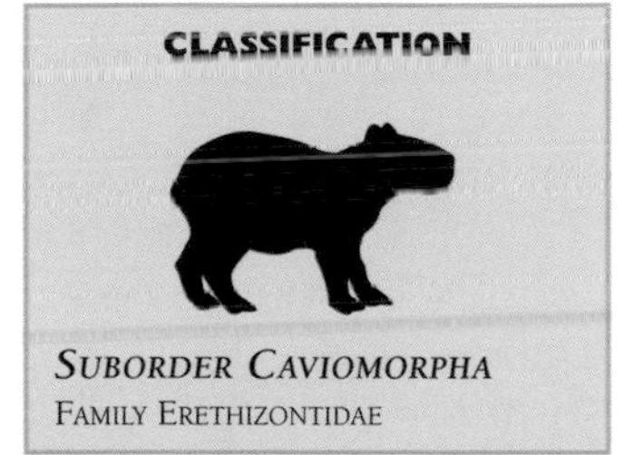

SUBORDER CAVIOMORPHA
FAMILY ERETHIZONTIDAE

Characteristics The prehensile-tailed porcupine is 11–24 inches (28–60 cm) long with a tail of 13–18 inches (33–45 cm). The quills are easily detached and remain embedded in an attacker by means of small barbs along much of their length. New World porcupines make up for their poor eyesight with keen senses of touch, hearing and smell.

Food Mainly nocturnal, the prehensile-tailed porcupine feeds on leaves, stems and some fruits.

Young A single well-developed offspring, rarely twins, is produced weighing about 14 ounces (400 g).

COENDU PREHENSILIS

The prehensile-tailed, tree or Brazilian porcupine is covered with short, thick spines, except for its tail, which is equipped with muscles and is used to grasp branches when the animal is feeding. The hands and feet bear long, curved claws on each digit, which assist it to climb with ease.

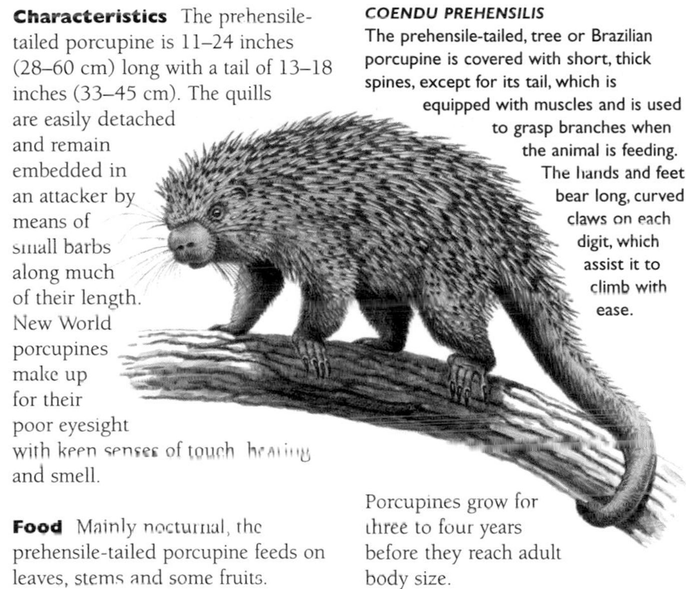

Porcupines grow for three to four years before they reach adult body size.

Habitat Prehensile-tailed porcupines occur in the forests of Bolivia, Brazil and Venezuela.

MARA

Despite its name, the mara or Patagonian hare is a rodent not a lagomorph. Its body resembles other cavies, but it has long rabbit-like legs. It is a social animal and lives in burrows, which it leaves to bask in the sun.

Close relatives The genus *Dolichotis* contains two species. Maras are related to other cavies, which include the domestic guinea pig.

Characteristics This animal is 28–30 inches (70–77 cm) long. Its tail is insignificant, a mere $1\frac{1}{2}$ inches (4 cm). The mara's long, slender legs are capable of propelling the animals at speeds of up to 28 mph (45 km/h).

DOLICHOTIS PATAGONUM
The mara is a monogamous species and pairs mate for life. During the breeding season, the male keeps watch while the female feeds, defending her against the approach of other males. Young bachelor males often "cradle snatch" by attaching themselves to infant females.

Food Maras feed primarily on short grasses and herbs.

Young Females give birth to a maximum of three well-developed young in a communal creche. Communal suckling does not take place. The female nurses her own pups for an hour or more once or twice a day for up to four months.

Habitat The mara occurs in the open scrub deserts and grasslands of central and southern Argentina.

Plains Viscacha

The plains viscacha lives in communal burrows, piling bones, stones, sticks and other objects around the burrow entrances. This agile, fast rodent is distinguished by black and white stripes across the face.

Close relatives Viscachas are related to chinchillas. There are six species of Chinchillidae in three genera.

Characteristics These expert burrowers dig mainly with their forefeet, pushing the soil off with their noses, the nostrils closing to prevent soil getting in. Their teeth are divided into transverse plates. They range from 18 to 26 inches (45–65 cm) long and have a black tail of 6–8 inches (15–20 cm). Males are larger than the females.

Food The plains viscacha grazes, especially on grasses also favored by domestic stock.

Young Females usually breed once a year, but two litters are sometimes born in mild climates. Two offspring is the norm.

Classification

Suborder Caviomorpha
Family Chinchillidae

Habitat The plains viscacha is found in grassland throughout Argentina.

Lagostomus maximus
Ranchers dislike viscachas because they compete with stock—10 viscachas eat as much as one sheep. Acidic viscacha urine destroys pasture and concealed underground burrows in the pampas are traps for horses and cattle, which are frequently injured by stumbling into them.

RED-RUMPED AGOUTI

Agoutis live in burrows among boulders or roots on the forest floor. These swift, shy animals are usually active in the daytime, but become nocturnal if their habitat is disturbed. They can jump up to $6\frac{1}{2}$ feet (2 m) off the ground vertically from a standing position.

Close relatives The six or so species of agoutis share the family Dasyproctidae with pacas and acuchis.

Characteristics Agoutis have four toes on their forefeet and three toes on their hindfeet. The claws are blunt and almost hooflike in some species. They weigh up to $4\frac{1}{2}$ pounds (2 kg) and their long, slender limbs are adapted for speed. The tail is vestigial.

Food Agoutis feed on fallen fruit, roots and leaves.

Young Agoutis produce two to four young after a gestation period of about three months.

Habitat Agoutis are distributed across the forests and savanna of Venezuela, east Brazil and the Lesser Antilles.

CLASSIFICATION

SUBORDER CAVIOMORPHA
FAMILY DASYPROCTIDAE

DASYPROCTA LEPORINA
The agouti is attracted to its favorite food—fallen fruit—by the sound of it hitting the ground. In times of scarcity, it buries some in burrows and thus plays a role in the dispersal of the seeds of many forest trees.

Capybara

The capybara is the largest species of rodent alive today. Its fossil record only goes back to the Pliocene epoch. These animals live in groups of about 15, which may temporarily join together in larger crowds.

Close relatives The capybara is the only living member of an apparently fairly recent family.

Characteristics The nostrils, eyes and ears of this animal are all high on the head, adaptations to its semi-aquatic life. There are partial webs between the digits of both hind and forefeet. Capybaras can weigh up to 145 pounds (66 kg) and their cheek teeth grow throughout life.

CLASSIFICATION

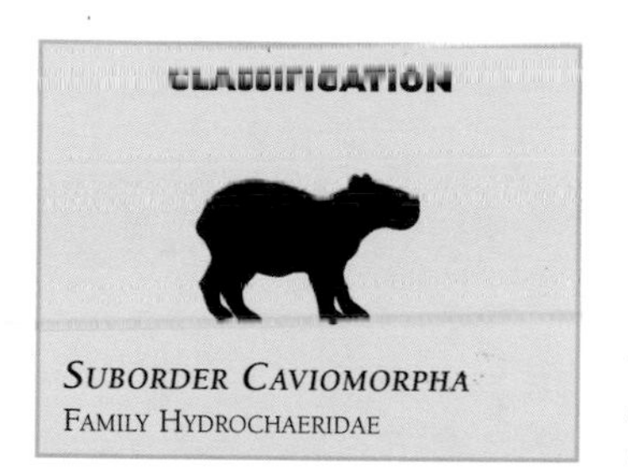

SUBORDER CAVIOMORPHA
FAMILY HYDROCHAERIDAE

Food At dawn and dusk, capybaras graze at the water's edge, and in cultivated areas take melons, grain and squash. They are much more efficient at digesting plant material than are cattle and horses.

Young The female produces a single litter of two to seven young each year on land. They are well-developed at birth, and when active, constantly emit a chirring purr.

HYDROCHAERUS HYDROCHAERIS

Capybaras take to the water to elude predators and to mate. In Venezuela, Roman Catholic missionary monks classified capybara meat as legitimate Lenten fare in the early sixteenth century. In some places, today, capybaras are cropped for their meat and valuable leather.

Habitat Capybaras are widespread, always near water, east of the Andes from Panama to Argentina.

KINDS OF MAMMALS

LAGOMORPHS

Two families comprise the order Lagomorpha: the diminutive pikas; and the long-eared rabbits and hares. The group are characterized by the two pairs of chisel-like upper incisors that bite against one pair in the lower jaw and a slit between the upper lip and the nostrils—the "hare lip." They also differ from rodents in having hair on the soles of their feet. Lagomorphs are familiar animals with often dense populations across a widespread range. The 14 species of pikas are the smallest lagomorphs, and have prominent rounded ears and no tail. There are about 65 species in the rabbit and hare family, which vary in their social behavior; reproduction; and burrowing and running abilities.

PIKA

The oldest known lagomorphs occurred in the late Eocene epoch (50 million years ago) in Asia and North America. Although most species of pikas resemble rodents, in particular the domestic guinea pig and hamsters, they are more closely related to rabbits and hares.

Close relatives
There are 14 species in the genus *Ochotona*.

Characteristics Pikas, which are smaller than rabbits, have short rounded bodies and no visible tail. They are sometimes called whistling hares or piping hares for the sounds they make. The mating call is a 30-second series of squeaks; everyday squeaks are much shorter.

Food The pika is an opportunistic vegetarian and will eat any available plant material near its mountainous home.

Young Pikas produce several litters of two to six furless young in the spring and summer after a gestation period of about one month.

Habitat This species inhabits rocky mountainous areas in western North America.

CLASSIFICATION

ORDER LAGOMORPHA
FAMILY OCHOTONIDAE

OCHOTONA PRINCEPS
In late summer and fall, pikas harvest vegetation, dry it in the sun and store it under overhanging rocks or in burrows. These hoards of hay provide winter fodder. Pairs of North American pikas defend their territory aggressively against trespassing neighbors.

Eastern Cottontail Rabbit

Rabbits have many enemies, including humans, who have hunted and eaten them for centuries. Rabbits rarely make any vocal sounds unless captured. They thump their feet in alarm or aggression both above and below ground.

Close relatives There are 13 species of cottontail rabbits, named for the white on the underside of their tails.

Lagomorph distribution

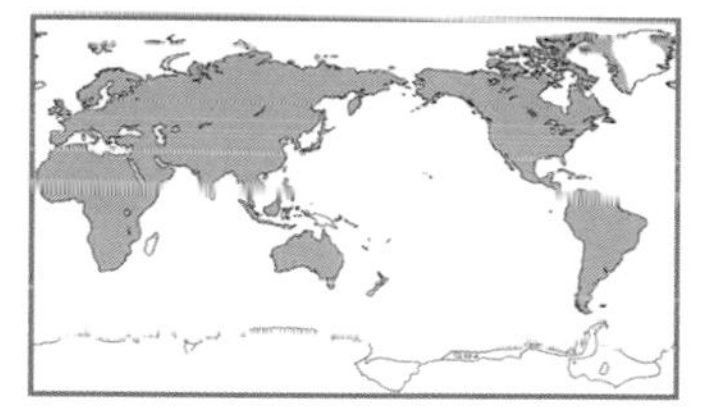

Widely distributed, either naturally or as a result of introduction by humans, lagomorphs are, however, absent from Antarctica and parts of South-East Asia, including several of the larger islands.

SYLVILAGUS FLORIDANUS

The eastern cottontail or Florida rabbit is a very common species, which breeds prolifically, and lives in communities, usually in open country. It is non-territorial and does not dig a burrow, but occupies those made by other animals or shelters in vegetation.

Characteristics Rabbits are generally smaller than hares, but like all lagomorphs, have long incisor teeth in the upper jaw that are constantly growing. Eastern cottontails are around 10–18 inches (25–45 cm) long and weigh from 1 to 4 pounds (500 g–2 kg).

Food The eastern cottontail is a herbivore that thrives mainly on plant material.

Young Does from early spring litters are fertile at about 10 weeks old. The young—blind and with only a sparse covering of hair— are born after 26–28 days' gestation.

Habitat The eastern cottontail occupies a wide variety of habitats, ranging over the entire USA, east of the Rockies.

Black-tailed Jackrabbit

The black-tailed jackrabbit is a desert hare. It is active at night and rests in the shade during daytime or digs short burrows to avoid high summer temperatures. This species tends to run when startled rather than to take cover. It may attain speeds of up to 35 mph (56 km/h) over short distances.

Close relatives There are about 65 species of leporids.

LEPUS CALIFORNICUS
The black-tailed jackrabbit's ears are its most distinctive feature and an effective temperature control for its hot, arid environment. When the blood vessels are engorged with blood, the animal loses heat; when they are constricted, body heat is conserved.

Characteristics The head and body length of the black-tailed jackrabbit is 18–25 inches (45–63 cm). Long hindlegs are adapted to its fast, bounding gait. In common with most species of hares, its ears are longer than its head.

Food Its preference for succulent green plants makes the black-tailed jackrabbit a pest to cultivated crops. In winter it eats more woody plants. Like all lagomorphs, jackrabbits ingest their fecal pellets for additional nutrients.

Young Females may produce several litters a year with one to six offspring in each. Leverets are born in a shallow depression in the ground, shaped by the animal's body and called a form. They are fully furred with their eyes open and can hop a few minutes after birth.

Habitat The black-tailed jack-rabbit lives on prairies, in arid scrub and cultivated areas of the USA.

CLASSIFICATION

Order Lagomorpha
Family Leporidae

Snowshoe Hare

This animal has been trapped for its valuable pelt for more than two centuries. In most places, its gray-brown summer coat turns pure white in winter. Like Arctic hares, snowshoe hares burrow into snow.

LEPUS AMERICANUS

The numbers of snowshoe hares in some populations fluctuate markedly on a rough 10-year cycle. This appears to relate to the abundance and quality of food, especially in winter. Birds of prey, lynxes, foxes, coyotes, martens, fishers and other predatory mammals all hunt snowshoe hares.

Close relatives Snowshoe hares are one of 19 species in the genus.

Characteristics The snowshoe hare has a head and body length of 14–20 inches (35–50 cm) with a short tail. The lower surface of its feet are well covered with long, thick, brush-like hairs that cushion the animal's movement on hard ground and give it grip when running on snow.

Food In summer, the snowshoe hare eats juicy herbaceous plants and grass; its winter fare is mainly twigs and bark.

Young Breeding begins in spring and a female may have two to three litters a year with usually four offspring. The leverets move to separate forms about three days after birth, but meet their mother daily around sunset for less than three minutes' suckling.

Habitat This species inhabits the forests, swamps and thickets of Alaska, Canada and the northern USA. It needs dense cover where animals hide during the day, emerging to feed only after dark.

Arctic Hare

This species is also known as the alpine hare or varying hare because, with the exception of populations in Ireland, it changes its dark brown summer coat for a white winter coat; only the eartips remain dark. The Arctic hare is one of the few species of hare to dig a burrow into snow or earth. This is used as a bolt-hole for the young and is rarely entered by the adults.

Close relatives Arctic hares are specifically related to other species of hares in the *Lepus* genus.

Characteristics Arctic hares prefer forested areas to open country. They are solitary animals and form pairs or small groups only in the breeding season. They are a largish hare and weigh up to 11 pounds (5 kg).

Food Arctic hares eat mainly bark in winter and have a more varied vegetarian diet in the other seasons.

Young Litters are born well furred, open-eyed and mobile.

Habitat In Europe, the Arctic hare occupies alpine and Arctic tundra and adjacent woodlands. In Canada and Greenland, it occurs only in the Arctic tundra above the treeline.

LEPUS TIMIDUS

A hare's long legs, especially the back ones, enable it to run from predators and to change direction speedily—a useful tactic when pursued by such birds as hawks, which cannot twist and turn as quickly. Nevertheless, many Arctic hares are killed by carnivorous birds and mammals.

CLASSIFICATION

ORDER LAGOMORPHA

FAMILY LEPORIDAE

European Rabbit

The European or common rabbit originated on the Iberian Peninsula and north-west Africa, and was only introduced into the rest of western Europe about 2,000 years ago. The ancient Romans were probably the first to domesticate the rabbit and all strains of domestic rabbit are descended from this species.

Close relatives The European rabbit is monotypic in its genus.

Characteristics Lagomorphs' eyes are large and adapted to activity at twilight and after dark. Rabbits have shorter ears and weaker hindlimbs than hares, and are scamperers and bounders that do not venture far from cover. The European rabbit is 13–18 inches (33–45 cm) long.

Food European rabbits mainly eat grass and leafy plants but can do serious damage to vegetable and grain crops and to young trees.

Young The European rabbit's fecundity is legendary. Does may produce three to five litters a year, averaging five or six young, though up to 12 have been recorded. The kittens are born blind and without fur and do not leave the burrow until they are about three weeks old.

Habitat This species is found throughout Europe and north-west Africa and has been introduced into many places including Chile, New Zealand and Australia.

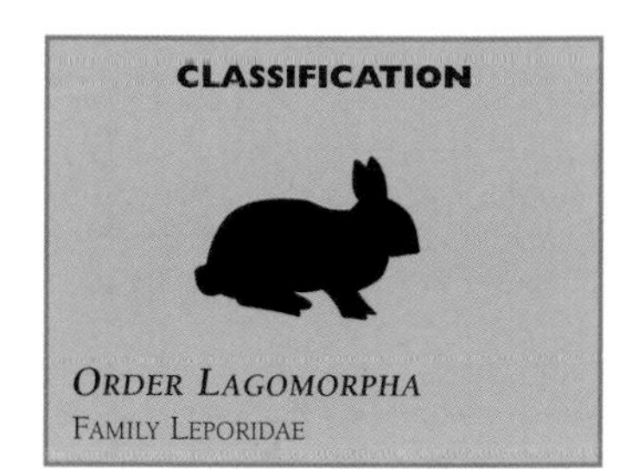

ORYCTOLAGUS CUNICULUS
European rabbits live in burrows called warrens, which protect them from the weather and from predators. Female rabbits use grass and fur plucked from their bellies to line special chambers for their young. Rabbits are usually silent but squeal when frightened or injured.

MAMMAL CLASSIFICATION

CLASS MAMMALIA	
SUBCLASS PROTOTHERIA	
ORDER MONOTREMATA	**MONOTREMES**
Tachyglossidae	Spiny anteaters
Ornithorhynchidae	Duck-billed platypus
SUBCLASS THERIA	
INFRACLASS METATHERIA	
ORDER MARSUPIALIA	**MARSUPIALS**
Didelphidae	American opossums
Microbiotheriidae	Colocolos
Caenolestidae	Shrew-opossums
Dasyuridae	Marsupial mice, etc.
Myrmecobiidae	Numbat
Thylacinidae	Thylacine
Notoryctidae	Marsupial mole
Peramelidae	Bandicoots
Peroryctidae	Spiny bandicoots
Vombatidae	Wombats
Phascolarctidae	Koala
Phalangeridae	Phalangers
Petauridae	Gliding phalangers
Pseudocheiridae	Ringtail possums
Burramyidae	Pygmy possums
Acrobatidae	Feathertails
Tarsipedidae	Honey possum
Macropodidae	Kangaroos, wallabies
Potoroidae	Bettongs
INFRACLASS EUTHERIA	
ORDER XENARTHRA	**ANTEATERS, SLOTHS & ARMADILLOS**
Myrmecophagidae	American anteaters
Bradypodidae	Three-toed sloths
Megalonychidae	Two-toed sloths
Dasypodidae	Armadillos
ORDER INSECTIVORA	**INSECTIVORES**
Solenodontidae	Solenodons
Tenrecidae	Tenrecs, otter shrews
Chrysochloridae	Golden moles
Erinaceidae	Hedgehogs, moonrats
Soricidae	Shrews
Talpidae	Moles, desmans
ORDER SCANDENTIA	**TREE SHREWS**
Tupaiidae	Tree shrews
ORDER DERMOPTERA	**FLYING LEMURS, COLUGOS**
Cynocephalidae	Flying lemurs, colugos
ORDER CHIROPTERA	**BATS**
Pteropodidae	Old World fruit bats
Rhinopomatidae	Mouse-tailed bats
Emballonuridae	Sheath-tailed bats
Craseonycteridae	Hog-nosed bat, bumblebee bat
Nycteridae	Slit-faced bats
Megadermatidae	False vampire bats
Rhinolophidae	Horseshoe bats
Hipposideridae	Old World leaf-nosed bats
Noctilionidae	Bulldog bats
Mormoopidae	Naked-backed bats
Phyllostomidae	New World leaf-nosed bats
Natalidae	Funnel-eared bats
Furipteridae	Smoky bats
Thyropteridae	Disc-winged bats
Myzopodidae	Old World sucker-footed bat
Vespertilionidae	Vespertilionid bats
Mystacinidae	New Zealand short-tailed bats
Molossidae	Free-tailed bats
ORDER PRIMATES	**PRIMATES**
Cheirogaleidae	Dwarf lemurs
Lemuridae	Large lemurs
Megaladapidae	Sportive lemurs
Indridae	Leaping lemurs
Daubentoniidae	Aye-aye
Loridae	Lorises, galagos
Tarsiidae	Tarsiers
Callitrichidae	Marmosets, tamarins
Cebidae	New World monkeys
Cercopithecidae	Old World monkeys
Hylobatidae	Gibbons
Hominidae	Apes, humans

Scientific name	Common name
ORDER CARNIVORA	**CARNIVORES**
Canidae	Dogs, foxes
Ursidae	Bears, pandas
Procyonidae	Racoons, etc.
Mustelidae	Weasels, etc.
Viverridae	Civets, etc.
Herpestidae	Mongooses
Hyaenidae	Hyenas
Felidae	Cats
Otariidae	Sealions
Odobenidae	Walrus
Phocidae	Seals
ORDER CETACEA	**WHALES, DOLPHINS**
Platanistidae	River dolphins
Delphinidae	Dolphins
Phocoenidae	Porpoises
Monodontidae	Narwhal, white whale
Physeteridae	Sperm whales
Ziphiidae	Beaked whales
Eschrichtiidae	Gray whale
Balaenopteridae	Rorquals
Balaenidae	Right whales
ORDER SIRENIA	**SEA COWS**
Dugongidae	Dugong
Trichechidae	Manatees
ORDER PROBOSCIDEA	**ELEPHANTS**
Elephantidae	Elephants
ORDER PERISSODACTYLA	**ODD-TOED UNGULATES**
Equidae	Horses
Tapiridae	Tapirs
Rhinocerotidae	Rhinoceroses
ORDER HYRACOIDEA	**HYRAXES**
Procaviidae	Hyraxes
ORDER TUBULIDENTATA	**AARDVARK**
Orycteropodidae	Aardvark
ORDER ARTIODACTYLA	**EVEN-TOED UNGULATES**
Suidae	Pigs
Tayassuidae	Peccaries
Hippopotamidae	Hippopotamuses
Camelidae	Camels, llamas
Tragulidae	Mouse deer
Moschidae	Musk deer
Cervidae	Deer
Giraffidae	Giraffe, okapi
Antilocapridae	Pronghorn
Bovidae	Cattle, antelopes, etc.
ORDER PHOLIDOTA	**PANGOLINS, SCALY ANTEATERS**
Manidae	Pangolins, scaly anteater
ORDER RODENTIA	**RODENTS**
Aplodontidae	Mountain beaver
Sciuridae	Squirrels, marmots, etc.
Geomyidae	Pocket gophers
Heteromyidae	Pocket mice
Castoridae	Beavers
Anomaluridae	Scaly tailed squirrels
Pedetidae	Spring hare
Muridae	Rats, mice, gerbils, etc.
Gliridae	Dormice
Seleviniidae	Desert dormouse
Zapodidae	Jumping mice
Dipodidae	Jerboas
Hystricidae	Old World porcupines
Erethizontidae	New World porcupines
Caviidae	Guinea pigs, etc.
Hydrochaeridae	Capybara
Dinomyidae	Pacarana
Dasyproctidae	Agoutis, pacas
Chinchillidae	Chinchillas, etc.
Capromyidae	Hutias, etc.
Myocastoridae	Coypu
Octodontidae	Degus, etc.
Ctenomyidae	Tuco-tucos
Abrocomidae	Chinchilla-rats
Echymidae	Spiny rats
Thryonomyidae	Cane rats
Petromyidae	African rock-rat
Bathyergidae	African mole-rats
Ctenodactylidae	Gundis
ORDER LAGOMORPHA	**LAGOMORPHS**
Ochotonidae	Pikas
Leporidae	Rabbits, hares
ORDER MACROSCELIDEA	**ELEPHANT SHREWS**
Macroscelididae	Elephant shrews

INDEX

Page references in *italics* indicate illustrations and photos.

INDEX continued

M

N

O

P

INDEX continued

T

U

V

INDEX

ACKNOWL

TEXT The text ... wrence G. Barnes, M. M. Bryden,
Carson Creagh, ... Tom Kemp, Judith E. King, Gordon
L. Kirkland Jr., ... nith, Jo Rudd (index), Jeheskel (Hezy)
Shoshani, Ian St ... a C. Sunquist, W. Chris Wozencraft
ILLUSTRATIONS ... Dogi, Simone End, Christer Eriksson,
Alan Ewart, Mi ... nson, David Kirshner, Frank Knight,
John Mac, Jame ... Barbara Rodanska, Trevor Ruth,
Claudia Saracen ... allace, Ann Winterbotham
PHOTOGRAPHS
CONSULTANT E ... f Biological Sciences, Macquarie
University, Sydn